A View from Above

The German pioneer Otto Lilienthal presented Professor George Francis Fitzgerald
with one of his gliders in 1895. Here the bearded professor is gazing at his machine in the grounds of Trinity College, Dub
He would race at great speed, to the cheers of his students but, unlike Lilienthal, he never got airborne!

A VIEW FROM ABOVE

Donal MacCarron

THE O'BRIEN PRESS
DUBLIN

This edition published 2000 by The O'Brien Press Ltd,
20 Victoria Road, Dublin 6, Ireland.
Tel. +353 1 4923333; Fax. +353 1 4922777
email: books@obrien.ie
website: www.obrien.ie

ISBN: 0-86278-662-2

British Library Cataloguing-in-Publication Data
MacCarron, Donal, 1927-
A view from above : 200 years of aviation in Ireland
1.Aeronautics - Ireland - History
2.Flight - Ireland - History
I.Title
623.1'3'009415

1 2 3 4 5 6 7 8 9 10
00 01 02 03 04 05 06 07 08

The O'Brien Press receives assistance from The Arts Council /
An Chomhairle Ealaíon

Layout and design: The O'Brien Press Ltd.
Colour separations: C&A Print Services Ltd.
Printing: Zure S.A.

CONTENTS

DEDICATION

This book is dedicated to the memory of
Tom Gilbert,
who first encouraged the author to look skywards,
long, long years ago.

The colour section in this book has been made possible by the
generous assistance of the Aviation Division of Irish Shell Ltd.,
Airbus Industrie, Toulouse, and Eurocopter SA, Paris.

ACKNOWLEDGEMENTS

In preparing this book I have had many pleasant conversations with many helpful people (some of them alas no longer with us), including: Comdt Jack Ryan BE DIC, Johnny Maher, Capt Jack Kelly-Rogers, Paddy Cummins (Waterford City), Dr Garret Fitzgerald, Bernard Share (Kildare), Grainne O'Malley and Bernie Kelly (Corporate Affairs, Aer Lingus), Garda Alice Kennedy (An Garda Press Office), Garda Joe O'Flynn (GASU), Madeleine O'Rourke, and many friends, past and present, in the Irish Air Corps. The stanzas that appear on pages 10 and 109 are from Geoffrey de Holden-Stone's poem 'Aircraft'.

My thanks also to my wife, Monique, who has endured the tantrums of an irritable writer!

For illustrations: I am indebted to Bob Montgomery (RIAC Guinness Segrave Archive), the Brendan Treacy Collection (Nenagh), Fergus Flood (Co. Limerick) for his father James's Foynes photograph, Meda Ryan (Ennis), Margaret O'Shaughnessy (Curator of Foynes Flying Boat Museum), George Flood (Dublin), Capt. Aidan Quigley, Danny Skeehan (Aer Lingus Studio Manager), Brenda Dwyer (Philatelic Bureau, An Post), Tony Flanagan (Air Corps Photo Section), Peter J Bish, the private collection of Comdt. E Kelly (via Military Archives), Robert Allen, Vincent Killowry (Ennis) for his excellent painting in the colour section, and John King, aviation photographer and writer of Auster Productions, Auckland 10, NZ, for his splendid cover photograph.

Front cover photograph: An Avro 631 Cadet training aircraft in its original Irish Air Corps colour scheme of 1934. This aircraft, with others of its type, served until the end of the 'Emergency' (WWII), before going through the hands of several civilian owners in Ireland. Eventually, a British Airways 'Concorde' captain acquired the aircraft, which was in very bad condition, and he handed it over to an expert aircraft restorer in the UK, but before the job was completed the airline captain retired to New Zealand and brought the Cadet with him. It was later acquired by the present proud owner, Jim Schmidt, who claims half-Irish ancestry, and was brought to flying condition by Smiths Tech-Air of Auckland. The restoration of this unique aeroplane was a major undertaking in time, effort and money, and has kept a piece of Irish aviation history alive.

Photograph by John King.

PREFACE

WHEN I WAS A YOUNG LAD in the early 1930s, playing on the beach at Laytown, I watched with great excitement as an aeroplane landed thereon. It can't have been a great landing because a splinter was knocked off the wooden propeller, or perhaps it had come adrift in the air and that was the cause of the landing. Whatever the reason, my playmates and I were delighted when the pilot disappeared to make a telephone call and we indulged in pushing the misfortunate aircraft to and fro along the strand. Some while later, a man in white overalls and the pilot refitted a new prop and, much to our annoyance, wrestled the broken one from us and stowed it in the aircraft. With adults marshalling us, the machine took off, rose and, becoming smaller and smaller, it was soon a dot in the sky before it disappeared altogether. This was my first experience of this phenomenon!

Returning home from holidays to the Midlands I didn't see another aeroplane for some years, but then I experienced a feast of flying when an air display came to visit the local racecourse at Mullingar. There were aeroplanes flying that had not only one, but two or even three engines – it was truly a world of wings! Biplanes were predominant but there was a couple of sleek modern monoplanes. These caused an old pilot of Great War vintage to remark to me: 'Dear boy, flying has never been the same since they took the top wing off!' I know exactly what he meant, though little did I think that one day I would be recording these halcyon days of flying and the exploits of all the Irish sky merchants from the beginning.

In this book I have attempted to pull together in chronological order all the happenings in Irish skies. I have purposely concentrated on what was once called the Irish Free State and is now the Republic. Apart from early Ulster pioneers like Lillian Bland and Harry Ferguson, I have not covered Northern Ireland, which has a unique aviation culture. As early as 1924, just south of Balmoral, the first municipal aerodrome in these islands was established as Malone Air Park, though it unfortunately lasted for only a year.

The only aircraft manufacturer in Ireland, Shorts Brothers plc, has long been established in Belfast, while during World War II the two peacetime aerodromes which existed in Northern Ireland grew to twenty and were the bases for high-level training and operational flying, particularly in the Battle of the Atlantic. All of this deserves a volume of its own. I the meantime I trust that I have covered the rest of the Irish sky adequately – any errors and omissions I claim as mine own!

Donal MacCarron, August 2000

I

Up, Up and Away!

IT SEEMS AS IF WE HUMANS HAVE ALWAYS WANTED TO FLY. Leonardo da Vinci, in the fifteenth century, was obsessed by the idea. He filled his notebooks with speculations on the mechanisms of bird flight and sketched designs of a variety of aenonautical devices. But no one succeeded in copying the birds and those who tried were either killed or injured. By the 1700s travelling showmen, who described themselves as 'flying men', had devised a method of gliding down to earth from high buildings using long ropes and balancing on a board. One such was a Mr Cadman, whose exhibition in Dublin was described thus: 'The famous

One of Leonardo da Vinci's concepts for a flying machine – not for the timid!

Flying Man flew twice, in a very extraordinary manner, from the top of King John's Castle in Thomas Street, to a post set up in Meath Street.' He came to a sticky end in 1740 and is commemorated on a tablet outside St Mary's Church in Shrewsbury: '… a faulty cord being drawn too tight, hurried his soul on high to take a flight, which bid his body here below, goodnight!'

At the beginning of the nineteenth century, Sir George Cayley discounted the idea of the flapping wings favoured by earlier experimenters and for the next fifty years he constructed fixed-wing gliders that made short flights with human passengers aboard. A stanza commemorates the exponents of the fixed wing:

> From Icarus to Bacon, we dared the deadly quest
> With courage never shaken by failure manifest,
> So strove to sail the glad air, so passed the torch alight,
> From Cayley unto Ader, to Langley, Roe and Wright.

But others disdained the idea of wings, and turned to the spherical balloon. In 1783 the French brothers Mongolfier demonstrated unmanned balloons filled with hot air, while another Frenchman sent a balloon heavenwards inflated with the newly-discovered hydrogen gas.

About this time the Marquis d'Arlandes built a tall tower on his estate in order to test his version of a parachute. He only succeeded in breaking his leg! Undeterred, the nobleman readily accepted the invitation of Jean-François Pilatre de Rozier to accompany him in the first manned balloon flight. They reached 3,000 feet and soared over Paris for about half an hour, to the great delight of the citizens.

February 1910: posing in the basket of the balloon St Louis is the wife of its owner, John Dunville, a founder member of the Irish Aero Club who, accompanied by Mr CW Pollock, flew it to Macclesfield, England.

In Ireland, people were kept well informed of all these developments by illustrated articles in the *Hibernian* magazine. In December 1784 the magazine recorded a total of twenty-seven balloon ascents, mainly French, up to that date. A year after the Mongolfier brothers' demonstrations, a Mr Riddick sent a balloon up from the Rotunda Gardens in Dublin, but he soon retired from aeronautics. He was succeeded by a young student, Richard Crosby of County Wicklow.

Richard Crosby was possibly the first to experiment seriously with ballooning in Ireland. On 19 January 1785 he ascended at Ranelagh in the view of 40,000, and stayed aloft for about fifteen minutes.

Crosby's early balloons were about 12ft in diameter and he filled them with hydrogen, which he made by pouring acid on scrap iron and zinc. In the summer of 1784 he was giving exhibitions in the Pleasure Gardens in Ranelagh for the enlightenment of the nobility, the gentry and the general public. One of his early balloons carried a cat and almost reached the Isle of Man.

Using the gate-money he had accumulated, Crosby built a much larger balloon in which he planned to cross the Irish Sea himself. At the beginning of 1785 he ascended before a crowd of more than 20,000 and flew over Dublin Bay, landing at Clontarf. He chose the Royal Barracks (later to become Collins Barracks and now a part of the National Museum) as the site of his next ascent, but he failed to get airborne.

A mezzotint shows 'The Bauld McGwire' being lifted from the sea. The patriot, Lord Edward Fitzgerald, is shown gesticulating on the right, though in fact he greeted the soon-to-be-knighted balloonist when he disembarked at Howth.

Undaunted, he disposed of every unnecessary item of weight, including himself, and persuaded a twenty-one-year-old Trinity College student to take his place. On this flight the valve rope jammed and the adventurous undergraduate had to pierce the gasbag, hoping to descend on Ireland's Eye. However, the balloon flew on before he eventually ditched, nine miles northeast of Howth Head. After half an hour in the water he was picked up, none the worse for wear. When he disembarked from a fishing boat at Howth, the Lord Lieutenant and Lord Edward Fitzgerald greeted him as a hero, and the former saw to it that 'the bauld McGwire' received a knighthood for his bravery.

Crosby was undeterred by failure, and others were now getting in on the act, to the extent that the Lord Mayor forbade further attempts, mainly because of absenteeism by workers whenever a flight occurred. Nevertheless, Crosby was airborne again from Leinster Lawn and, despite colliding with a wall and killing three onlookers, he managed to reach a height from which he could view the Welsh coast. The flight was short-lived; Crosby was swept down into the sea and dragged along so fast that a rescuing boat could not get to him for two hours. However, this intrepid aeronaut was not going to be bested. Giving Dublin a miss, he opted for a Limerick take-off. He reached a height of 20,000ft over Ballygirreen in Clare, and he descended at Ennis, none the worse for having no oxygen – normally essential after the 10,000ft mark. Crosby never did attain his goal of crossing the Irish Sea, but he had started a craze for ballooning in Ireland.

As the sport developed it brought its own hazards to the surrounding environment. Because fire braziers were used beneath unmanned small balloons, considerable damage could be caused when they alighted on thatched roofs. In Tullamore, the town had to be virtually rebuilt after almost 100 houses were destroyed by such conflagrations!

Officialdom put its foot down firmly against such dangerous pursuits, and a generation was to pass before the Sadlers, father and son, attempted to conquer the Irish Sea. James, the father, had made several successful hot-air balloon flights, though on occasion he had ended up in the sea. On 1 October 1810, Sadler the elder took off from the Rotunda Gardens. Passing over Howth Head he discovered a hole in the balloon, which he staunched with his cravat. Soon he was over the Isle of Man and heading for Liverpool. A westerly airstream took him to Holyhead, but here a southerly breeze pushed the balloon back over the sea and down to another water landing! There were many ships in the sea below him and he called to one to spike his balloon with its bowsprit, and thus did Sadler and his balloon arrive eventually at Liverpool.

Where the father had failed, the son, Wyndham Wallace Sadler, succeeded. He took off in July 1817 from Portobello (now Cathal Brugha) Barracks and, waving his hat and flags and indeed drinking to the health of the 100,000 onlookers, he was on his merry way. After six hours in the air he landed safely in a cornfield close to Holyhead. The captain of the mailboat who lived nearby was the first to congratulate him. Sadler the younger completed other successful crossings but during his thirty-first ascent he was killed in a crash near Liverpool.

There were other balloon fanatics in Ireland. John Hodsman managed to lengthen the flight to the UK by getting as far as Westmoreland in his balloon 'The Raven', while a couple of months later a Joseph Simmons reversed the trend by flying from Preston in Lancashire to Bodenstown in Kildare. Like the space shuttles of today, balloon flights were becoming commonplace in the latter part of the nineteenth century. In addition, the 'gate' from the attendance, which had sometimes reached as many as 100,000 fans, was falling in comparison to expenses, which were steadily rising. Subsequent balloon flights were neither as long nor as historic as those of Crosby and the two Sadlers, though George Bernard Shaw became an enthusiast briefly at the turn of the century. To celebrate the success of his play *Man and Superman* in London, in 1907, Shaw went for a balloon flight, taking along the lead man, Robert Loraine, another actor, and his sister-in-law. The unromantic take-off point was Wandsworth Gasworks (which had, of course, the advantage of having plenty of gas on tap!), and according to Loraine:

> We were soon floating above the clouds at about 9,000 feet, exhilarated but somewhat awed by our first experience of altitude. After about forty minutes drifting – all very pleasant and seraphic with nothing happening except that Shaw *would* peer through a hole in the basket floor, which made him feel rather sick – we discussed landing. We wondered what our reception would be on coming down in someone's garden. I thought the people would be rather interested, and flattered when they discovered Shaw's identity. 'Don't be so certain,' said he, 'they may think my works are detestable!'

When they did land, a purple-faced individual rushed towards them, waving his shooting stick, and curtly instructed the bold aeronauts to quit his property. This form of reception occasionally occurs to this day!

The first modern hot air balloon in Ireland was acquired by the Dublin Ballooning Club in the summer of 1968. It was an all-black Picard type balloon of 57,000 cubic feet capacity. This type was named after the pioneer Auguste Picard, who in 1931 was the first to enter the stratosphere in a balloon with a pressurised cabin. A distant relation, Don

Picard, instructed the early members of the Club and the sport was off to a flying start, initially in the skies of County Kildare. By the beginning of 1973 the Club had two balloons, which it took to the World Championship for hot air balloons in New Mexico. Since then, Club pilots have represented Ireland at many other international events, on one occasion reaching a height of 19,000 feet while covering a distance of 330km. On St Patrick's Day, 1976, the multicoloured balloon number three was acquired and was appropriately called 'Rainbow'. In the words of one of the Dublin Ballooning Club's founder pilots: 'After more than thirty years, hot-air ballooning in Ireland flourishes as a sport, and on balmy days there is no better way to see our beautiful countryside in silence and serenity.'

The aptly named Tar Baby ascends at an Irish Ballooning Club meet at Abbeyshrule.

II

EARLY BIRDS

IN THE LATER YEARS OF THE NINETEENTH CENTURY, the greatest social changes, at least in Dublin, were probably in transport. To the ubiquitous strength of the horse, the steam power of the first railways was added, and later that mysterious power, electricity, lit the streets and drove its tramcars. 'What is the world coming to at all?' would have been a common question on the lips of citizens during the Edwardian era, prior to the Great War. The appearance of 'early birds' – manned, powered aircraft – would have made an impression at least as great as the ballooning activities of the preceding two centuries. As Ireland looked skywards, the question re-echoed from the citizenry: 'What is the world coming to, at all, at all?'

Though balloon fanatics had managed to defy gravity and take to the air, they were still dependent on the whims of wind and weather. Aeronautical enthusiasts yearned for a machine that would bring them closer to the birds. It had been shown fairly conclusively that flapping wings would not work, and so experiments with fixed-wing gliders offered a step in the right direction. Eventually the power of the internal combustion engine, like that of the 'horseless carriage' which ushered in the age of the motor car, was harnessed in the pursuit of the freedom of the air.

Across the Atlantic in North Carolina, two bicycle mechanics, the brothers Orville and Wilbur Wright, approached the problems of flight in a very scientific manner. By 1903 they had gained extensive knowledge of the aerodynamics of flight, having developed a series of powerless machines. They had also developed a lightweight engine, suitable for aviation. Combining both elements, they achieved the world's first sustained, controlled flight in a heavier-than-air machine. Thus the sky was fully conquered in the early years of the twentieth century.

Back in Ireland, just six years after the success of the Wright brothers, a young Belfast garage mechanic achieved the same result. Harry Ferguson's plane differed from the Wrights' in many respects – it was a monoplane as distinct from a biplane, and its engine was in the nose rather than behind the wings, pulling rather than pushing it through the air. Initially, Ferguson achieved heights of about twelve feet – it could hardly be called altitude! He continued his experiments until he could fly with his mechanic, Joe Martin, for distances of over three miles. It was claimed that Martin was Ireland's first aviation passenger, though, as we shall see, there was another contender for this distinction. After two more years Harry Ferguson turned away from flying and eventually developed the Ford Ferguson tractor. His tractors were made in Cork and by the late 1940s had virtually displaced the plough-horse on Irish farms.

Concurrently with Ferguson's endeavours, though quite separately, a young woman named Lillian Bland was also helping to get Ulster into the air. She had received a picture postcard of Louis Bleriot's monoplane, complete with its dimensions. Her home had a well-equipped workshop where she constructed a model biplane which flew with success as a kite. Spurred on, she made a full-sized glider with the intention, if it worked, of adding an engine later.

Early in 1910 she completed the aircraft. It had a 27ft 6in span and she christened it *Mayfly*, perhaps due to some slight doubt about its life expectancy! *Mayfly* soared well into the wind, though weak spots were shown up in the airframe which required strengthening. Lillian Bland acquired the help of four members of the Royal Irish Constabulary and a neighbouring boy to test the aircraft's capacity for taking the weight of an engine. The five held onto the machine as it rose abruptly. The constables then dropped off smartly, leaving the boy to hang on and bring it to earth. Lillian now contacted Alliot V Roe (the pioneer aviator and founder of Avro – AV Roe and Co., Ltd, the great British aircraft manufacturing firm), who produced a light, two-stroke air-cooled power unit. However, in a bench test, as the engine developed 20hp at 1,000 revolutions its propeller disintegrated, fortunately missing the onlookers. Roe supplied a new adjustable-pitch propeller and all was well.

Lillian Bland photographed one of her helpers at the controls of the powered version of Mayfly.
Like its predecessor, this home-built craft was structurally 'out of true'.

Mayfly didn't have a petrol tank as yet, so for the first powered try-out a whiskey bottle was used to feed the engine, via an ear-trumpet belonging to Lillian's elderly aunt. In the late summer of 1910, Lillian Bland was successfully flying Ireland's first powered biplane. After modifications to *Mayfly I*, this intrepid young woman built a larger version which she advertised in *Flight* magazine for £250, but purchasers had to provide their own engine – an early version of 'No Batteries Supplied'! A model of *Mayfly* is on display in the Ulster Transport Museum, as is a full-scale replica of Ferguson's monoplane, constructed by the noted Irish flying boat captain, Jack Kelly Rodgers.

In 1912 at his garage in Portlaoise, known in colonial times as 'Maryborough', a Mr Aldritt built a monoplane and took to the air while his wife gazed on, terror-stricken at his antics. A young local lad, when not under tutelage by the Christian Brothers, had assisted in the construction of the plane. He was James Fitzmaurice, and this experience sparked off an enormous enthusiasm in him for aviation. His career in the air reached its peak when he shared the piloting of the first successful east–west crossing of the Atlantic, sixteen years later. The local clergy of both denominations condemned Aldritt's experiments, declaring: 'It is unnatural for man to fly'! (Luckily, the clergy have never stopped people from swimming in the absence of human fins!) Aldritt's aircraft eventually came to grief and its remains were consigned to the rafters of his garage. It is interesting to note that apart from a few engine components, all its parts were *déanta in Éireann* (made in Ireland).

Progress could not be halted, and members of the Irish Automobile Club were destiny bound to establish the Irish Aero Club, which staged a two-day display at Leopardstown Racecourse in August 1910. Three pilots shipped their machines over from England, and these were housed in special hangars whose construction involved the removal of barrier rails and telegraph wires. Uncertain weather conditions did nothing to deter the crowds which, the newspapers reported, 'could only be matched by the Horse Show'. What a day it was! Virtually all of Dublin journeyed out

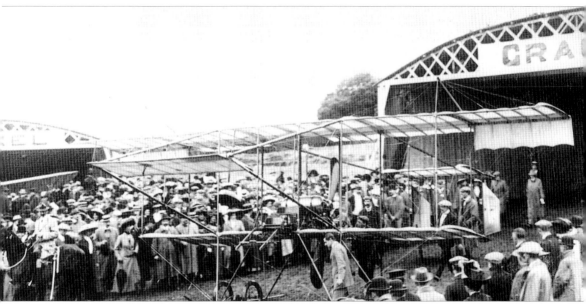

The Leopardstown meeting, August 1910. Cecil Grace's Farman biplane is prepared for flight outside one of the specially constructed hangars. In the background can be seen the similar hangar erected for Drexel.

by train from Harcourt Street Station. There were plenty of gatecrashers among them, people who, to avoid the half-crown entrance fee, climbed onto roofs and trees in order to get a good view. The gentry paraded in the

Flanked by two hussars, Drexel's Bleriot monoplane
is pulled into position in readiness for a flight.

members' enclosure, decked out in their finery, while the ordinary people were kept in their places by mounted hussars. Entertainment was provided in the general enclosure by the music of the Band of the Rifle Brigade. The visiting airmen, Messrs Grace, Drexel and Dickson, kept up a thrilling display, despite the necessity for occasional repairs. One of the machines was a two-seater and Desmond Arthur of Ennis was given a 'flip' at a cost of £10, after which he claimed that *he* was now the first aerial passenger in Ireland, no matter what Harry Ferguson's mechanic would have to say! The Leopardstown meeting produced a healthy balance of £421 on the credit side for its organisers.

The new club now decided to organise an air race from Dublin to Belfast and back, and this eventually took place in 1912. One of the motives for the race was to allow spectators along the route between the two cities to participate. Twelve competitors were to compete for prizes totalling over £300. However, for a variety of reasons, only four took off. The 'list of runners' was as follows: Henry Astley in a 70hp Bleriot; Lieutenant JC Porte, an ex-submariner, in a 100hp Deperdussin; James Valentine in a 50hp Deperdussin; and the Clareman, Desmond Arthur, who had been bitten by the flying bug at the Leopardstown display and now piloted his own machine, a 70hp Bristol.

Conditions could not have been worse, with gales, gusting winds, heavy rain and cloud persisting until 4.30 in the afternoon, when the competitors were at last able to depart. Because of the late start it was decided that the race would finish at Belfast, as darkness would overtake the return leg. Henry Astley and James Valentine got as far as Newry; Desmond Arthur failed to get airborne and damaged his machine by colliding with a flagpole; Lieutenant Porte (who had forsaken underwater craft for a higher calling) could not keep his craft on an even keel, because the gusting wind was pulling the control stick from his hands, and he was forced to return to the Racecourse.

Henry Astley takes off from Leopardstown in his 70hp Bleriot in the great Dublin/Belfast air race – but neither he nor any other competitor reached their destination!

Up at the Balmoral Showgrounds in Belfast the crowds were bitterly disappointed, having seen no flying whatsoever. To compensate for this a display was organised there but, tragically, Henry Astley was killed during it. His engine cut out and, as he turned to avoid the crowd, his powerless machine side-slipped into the ground. James Valentine stayed on in Ireland, carrying out demonstration flights across the width of the country. Desmond Arthur was commissioned as a lieutenant in the newly-established Royal Flying Corps, but was killed while practising at its Montrose base, thus becoming the first RFC fatality. The tuberculosis which had caused Lieutenant Porte to abandon submarines and take to the air eventually killed him, but not before he had carried out some very important work on the design and development of flying boats.

As for the Irish Aero Club, the outbreak of the Great War and the departure of several of its members to join the Royal Flying Corps caused its demise after only five years. It donated its surplus funds to the Red Cross. However, a decade later, its successor arose phoenix-like from the ashes of the world conflagration.

Robert Loraine, the actor who had played the lead in many of the dramas of George Bernard Shaw, had forsaken ballooning and taken to the air in a Henry Farman biplane. He was very keen to be the first across the Irish Sea and in August 1910 he took off from Holyhead. He climbed to 4,000 feet but it soon became evident that, after a previous minor landing accident, the plane had not been properly rigged. He began to hear the 'ping, ping' of the flying wires which held his plane together beginning to snap. As if this was not enough, his Gnome rotary engine stopped, but when he was down to 2,500 feet it restarted. Five times this happened before he sighted the Irish coast, and on each occasion he had to dive to clear the fuel starvation that was the cause of the problem. The trouble was that every time he dived another few wires would snap, and it was

In 1910, actor and aviator Robert Lorraine takes off from Anglesea in his Farman en route to Ireland. Unfortunately, he ended up in the sea just a quarter of a mile off Howth Head.

very evidently only a question of time before the whole machine would fold up in the air. Once across the coast at Howth he decided to get down as quickly as possible rather than keep going to the Phoenix Park as he had planned. However, just as he turned to land by the Bayley Lighthouse, the engine and the aircraft gave up the ghost and he landed about a quarter of a mile out to sea. He swam for the shore, but was picked up by a packet steamer. Ironically, the play in which he had been appearing in London was entitled *The Man from the Sea*!

A couple of years later, Denys Corbett Wilson attempted the crossing but chose a narrower stretch of the Irish Sea, St George's Channel. Corbett Wilson, whose maternal grandmother was from County Wicklow, lived the life of a country gentleman in Jenkinstown, County Kilkenny, with his widowed mother. Despite a wound in his leg, suffered in the Boer War, he was a keen sportsman. He had learned how to fly at the Bleriot School in southern France and was sufficiently well-off to indulge in what was basically a 'gentleman's pursuit'. A friend, Leslie Damer Allen of Limerick (a naval architect in London), also wished to make the crossing to Ireland and they decided to set off together from Hendon Aerodrome, near London, to see who would be the faster. They each owned a Bleriot monoplane Model XI, powered by a 50hp engine, so the odds were even. These machines were of such light construction that a headwind of more than 20mph would leave them standing in the air.

Both aviators got under way on 17 April 1912, and after several stops Corbett Wilson got as far as Radnor in Wales. He started his sea crossing at 6 o'clock in the morning and everything went smoothly until, about twenty miles out from the Wexford coast, he flew into a storm of rain, hail and heavy cloud. His only navigational aid, a compass, went berserk but the storm cleared and he got down safely into a field at Crane, near Enniscorthy. He had flown seventy miles at a height of 600 feet in one-and-a-half hours. His friend and rival, Leslie Damer Allen, did not survive the race across the Irish Sea and no trace was ever found of the unfortunate man or his aircraft – like Corbett Wilson, he had not troubled to equip himself with a life jacket. A few days after Corbett Wilson's crossing in 1910, Captain Vivian Hewitt made the longer crossing between Holyhead and Dublin, and reached the Phoenix Park in one-and-a-quarter hours.

Corbett Wilson succeeded again in another sea crossing, and was a well-known hero who drew crowds whenever he demonstrated his aerial skills at sports days and agricultural shows. When the Great War broke out he was a member of Number 3 Squadron of the Royal Flying Corps. By an odd coincidence, he was accompanied by the actor and fellow aviator Robert Loraine on a mission, when the actor was hit in the lung by ground fire. Corbett Wilson got him back to base swiftly and medical treatment saved his life. Corbett Wilson's own luck ran out over the Western Front after ten months, when he was struck by a German anti- aircraft shell and perished.

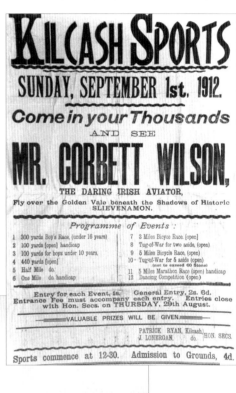

Above right: A poster shows that Denys Corbett Wilson was the main attraction at this Tipperary Sports Day.

Above left: Denys Corbett Wilson poses in front of his Bleriot XI before one of his many displays in Munster.

Corkonians had their own aerial hero at this time, in the person of young Josh C Evans-Freke, the tenth Lord of Carbery, who preferred to be known simply as 'JC'. He possessed a French Morane Parasol F, in which he performed the first loop-the-loop in Ireland, in July 1914 at the Clonakilty Agriculture Society's annual show. A reporter from the *Cork Examiner* described the event:

> Amidst breathless silence, the machine darted forward. It ran along the ground for about fifty yards and then slowly and very gradually rose in the air against a fairly strong south-west breeze. As the machine neared the boundary of the grounds the crowd became tremulous with fear for the safety of the aviator as it appeared as if he and his machine would collide with the corrugated iron fence. The suspense was only momentary, however, for the machine rose sharply and gracefully amidst the loud cheers of the crowd. Having performed two loops he then executed the 'falling leaf' descent before flying away.

A few days later, JC was winging his way to Paris as one of the six competitors in the famous London-to-Paris air race, but on the return leg he crashed into the sea. A trawler fished him out and transferred him to HMS *St Vincent* where he was soon drinking gins in the officers' wardroom. During the Great War, which erupted a few days later, Lord Carbery flew fighters. At the end of the war he renounced his title and all his lands in Ireland and moved to Kenya where, on his huge coffee plantation, he had a private airstrip, complete with hangars and a full-time maintenance crew.

LORD CARBERY.

-o&&&-&-&&o-

University Football Grounds.

- -o&& &-& &&o-

THURSDAY, 9th JULY, 1914.

LANDONS, CORK.

On the eve of the Great War, 'the flying earl of Bandon' gave several displays in County Cork, including one in the city.

III

LAND, SEA AND SKY

There are no skies like Irish skies
There are no clouds like Irish clouds

George Bernard Shaw

IRELAND'S LAND, SEA AND SKY, significant factors in the development of aviation, are all results of an unusually complex meteorological and geographical history. The coastline extends for 35,000 miles, yet nowhere is the sea further than seventy miles away. The effects of the great sweep of the Atlantic and of channel currents are felt everywhere. Though contrary winds can be literally blown away by the power and altitude of jet aircraft, even today the approaches to some Irish airports give a feeling of being swept in from the sea – for example, the approach to Shannon International Airport from the Atlantic, coming in low over the great river's estuary and an ancient O'Brien castle. But early aviators had a lot more than the view to think about – they had to pay great respect to the kaleidoscopic temperament of Irish weather, typically characterised by lots of wind and rain.

The character of the Irish landscape is distinctive: there is less tillage in Ireland than in Europe generally, and most of the arable land is given over to pasture. One-sixth of the land is bog, and when machinery moved onto it in the 1940s, to feed the various turf-burning power stations, the appearance from the air of these parallel lines of machine-cut turf was that of well-camouflaged aerodromes. During those years, when turf was such a valuable resource, several warplanes in distress were initially confused and at least two of them made disastrous forced landings in bogs. These misjudgments hark back to 1919 when the

Vickers Vimy of Alcock and Brown was blown in by the prevailing north-westerly winds onto what appeared to be a solid green field. It wasn't; it was Derrygimlagh Bog!

In terms of the development of aviation, Ireland has benefited from its location on the edge of the Atlantic. Early aviators quickly became aware of the long 'runways' available for their waterborne craft. Valencia Island appealed to flying boat devotees as long ago as the mid-1920s. The energetic pioneer Frank Courtney considered it as a possible starting point for his attempt at the unconquered east–west Atlantic trail in his twin-engined Dornier flying boat. He eventually decided on the safer southerly route via the Azores, though in the end his attempt fizzled out in northern Spain.

Valencia was also considered by the Italian Air Minister General Balbo as a transatlantic jumping-off base. He had successfully led a formation of ten large flying boats across the South Atlantic in 1930, and was now ready to tackle the North Atlantic with a mass formation of twenty-four seaplanes. Valencia looked like an appropriate stopover on

Valencia, County Kerry, 1931. Irish and Italian air officers discuss a possible interim landing place for an armada of Italian flying boats on a transatlantic flight.

the return journey, and a team of Italian and Irish Air Corps officers carried out a detailed survey there. However, the Italians eventually decided to come home via the Azores.

Two of the early Irish Air Corps machines, the Martinsyde Passenger (*The Big Fella*) in 1922 and a Fairey IIIF in 1928, were supplied complete with floats, but 'off-water' operations were never practised. Seaplanes, or amphibians, which can operate off land or water, were a priority from a military standpoint. Colonel (later Lieutenant General) Costello visited the United States on a purchasing mission in 1939. He was convinced of the value, in war conditions, of planes that could operate from inlets on Ireland's southern coastline and from lakes and dammed rivers. High-performance aircraft were not essential, as reconnaissance would be their main employment. The planned coastwatching service would also need the support of aircraft that could land beside suspect vessels.

During his visit, senior US Navy officers assured Colonel Costello that one squadron, or possibly two, of Grumman amphibians could be made available at the right price. However, there was strong opposition from both the Department of Defence and the Army Quartermaster's Branch and the offer was not taken up. Though some amphibians and supporting workboats were bought from Britain

General Costello's proposition that lakes, rivers and canals could offer versatile bases for waterborne aircraft during World War II is exemplified by this civilian Cessna Skyhawk II, EI-CFP, moored in the shelter of a bridge over the Royal Canal.

in the spring of 1939, that source dried up with the onset of war. The shortage of supply eventually meant that the Irish amphibians could only operate from land, as no spare wing-floats were available. A maritime dimension to the Irish military was provided solely by the hastily-organised Marine Service. Sixty years on, ship-borne helicopters and a useful naval service reflect a change of heart and the realisation that we are, in fact, surrounded by water! The British Royal Air Force kept flying boats and seaplanes on its inventory right up to the 1950s and the jet era – and even used them during the Berlin Airlift.

Air Corps maritime patrols keep a watchful eye on the country's valuable sea resources.

Under the terms of the Anglo-Irish Treaty of 1921, Britain retained landing rights in the Irish Free State, in the vicinity of the so-called 'treaty ports', until they were handed back in 1938. These rights were rarely exercised and a goodwill visit in the summer of 1930 was not considered to be political in any way. On this occasion, three Supermarine Southampton flying boats from 204 Squadron based at Plymouth visited Dún Laoghaire, made perfect landings, and entered the harbour in single file. The crews came ashore and were welcomed on the steps of the Royal St George Yacht Club by Captain Folsberry Holmes, the harbour master of what he persisted in calling Kingstown Harbour! The commander of the flight, Squadron Leader Lawrence, stressed that this was a friendly and unofficial visit for training purposes, and that all hands were very pleased to have the opportunity of visiting Dublin. He, along with some of his colleagues, was no stranger to Ireland – though he did not give details of his previous service here.

A few days after their arrival the flying boats took off for Cóbh, where the Royal Navy still had anchorages in what it called 'Queenstown Harbour'. The proponents of this great harbour as a future airbase were happy to witness a practical test of its facilities. The 'guard ship' based there was the destroyer HMS *Seawolf* and some of its sailors were taken up on sightseeing flights. After a week's stay, the flying boats, like great swans, took off for home. Eighteen months later, another pair visited Dún Laoghaire, where the commander felt at home, being a native of Sandycove. The next flying boat to alight at Dún Laoghaire was one of the Irish Air Corps' Supermarine Walrus amphibians at the end of its delivery flight in 1939. It had become separated from its two companions in extremely bad weather when en route from Southampton.

On 3 September 1939, a fateful day that year, an RAF Lerwick flying boat, distressed by the weather, alighted near the harbour. Its crew was unaware that World War II had broken out at eleven o'clock that morning, and so their flying boat unwittingly became the first of many 'belligerent' aircraft to find a haven in Ireland.

IV

THE MAGNIFICENT AIRBORNE SEVEN

For I dipt into the future, as far as human eye could see,
Saw the Vision of the world, and all the wonder that would be;
Saw the heavens fill with commerce, argosies of magic sails,
Pilots of the purple twilight, dropping down with costly bales;
Heard the heavens fill with shouting, and there rain'd a ghastly dew;
From the nations' airy navies grappling in the central blue.

ALFRED LORD TENNYSON'S VISION of commercial and military aviation, penned in 1854, came true, though in the early days of powered flight the aeroplane's potential as a weapon was only dimly perceived and often discarded. 'A useless and expensive fad, advocated by a few individuals whose ideas are unworthy of attention,' was the opinion of General Sir WG Nicholson, Chief of the British Imperial General Staff, in 1910. Two years later the British Army manouevres completely vindicated the Royal Flying Corps, formed on 13 April 1912. Some people still needed convincing, however. The first military exercises, which utilised aircraft for reconnaissance, caused a senior 'brass hat' to remark: 'The aeroplanes completely spoilt the war!'

At first, the main role of aircraft was seen as reconnaissance, but this, of course, could mean combat and experiments were carried out with rifles and light machine guns. Aeroplanes were not yet regarded as a substitute for traditional reconnaissance by cavalrymen, and the airmen had to demonstrate to the ground troops what aircraft could do for them.

Before the 'war to end all wars' broke out in the summer of 1914, the military potential of the aeroplane was being tested by Europe's major powers and, indeed, used to some effect in minor conflicts. Ireland became a proving ground for this new, and soon to become deadly, weapon.

A decade after the Wright brothers' breakthrough, 'flying machines' were now taking on an outward appearance that would remain largely unchanged for the next quarter of a century. Biplanes were to the forefront and their hitherto uncovered wood-and-wire bodies were now covered by unbleached linen, made taut with astringent dope.

At the beginning of 1913, Number 2 Squadron of the infant Royal Flying Corps (RFC) took up residence at their new base at Montrose in the east of Scotland. During the spring it was decided that seven of its aircraft would fly to Munster, where large-scale military exercises were planned by the British Army's Irish Command for September 1913, concurrent with similar manoeuvres elsewhere in the UK.

Six of the seven machines that were to leave Montrose were BE2as, a type which had evolved from a 1910 design. The BE2a was a two-seater powered by a 60hp Renault engine which gave a maximum speed of 70mph. The seventh aircraft was a French Farman 'Longhorn', a skeletal 'uncovered' primary trainer whose complex wooden structure obligingly absorbed much of the shock of pupils' efforts at landing. Its pilot, Captain George William Patrick Dawes, was the first serving officer of the British Army to be awarded an English Pilot Certificate. His brother, a lieutenant, was also one of the group. Dawes senior rose to the rank of General during the Great War and he was much decorated by his own country and many of its allies. He later made the first aeroplane flight in the Indian sub-continent. But all of that was in the future.

The Farman 'Longhorn's' fuselage was an open structure, unlike the other aircraft.

On the day before departure, one of the 'magnificent seven' was damaged and only six started out for Ireland – the first sea crossing ever undertaken by the Corps. One of these damaged its propeller on take-off and had to remain behind for repairs while yet another, well out over the sea, had to return when its flotation cylinder came loose. As a precaution

The aircraft that didn't get away to Rathbane, having crashed the day prior to take-off. The Magnificent Seven' were thus reduced to six.

against ending up in the sea, inflated cylindrical airbags had been mounted under the lower wings of all aircraft. However, the pilots complained that these caused the aircraft to be sluggish in getting off the ground and difficult to handle in the air during rough weather.

Initially, only four aircraft reached their destination – Rathbane, now subsumed by Limerick City. There were unplanned stops en route but eventually the two other stragglers caught up. A base camp had been established in a forty-acre field and when the pilots commenced practice flights up to 10,000 people gathered to watch. By the second week the exercises were in top gear, with 20,000 troops deployed over counties Limerick, Tipperary and Kilkenny. Some of the aircraft supported the

A pair of BE2a's get ready for the fray.

'Brown Army' and the others the 'White Army'. The weather was often inclement but the planes flew whenever possible, using any convenient field for delivering reports, or for forced landings. Whenever a plane came in to land, Army medical staff would stand by, underlying the perilous nature of flying in those days.

In Tipperary, there was a 'pitched battle' between the two armies which lasted five hours, resulting in the defeat of the Brown Army. A report from the battlefield stated:

> During the early stages of the fight, the airmen of both sides were constantly in flight. The amount of information the observers in these machines were able to convey to their respective generals either directly or by means of messages dropped from the dizzy heights to units of the force was most valuable. Owing to the favourable weather conditions and the clearness of the atmosphere, the aviators were easily able to rise above the 3,000-feet minimum height laid down by the regulations in order that they should be out of range of ground fire. As far as could be judged, many of them rose at times to an altitude considerably over this. All the while the machines went steadily, and fortunately without any accident to their intrepid pilots and observers.

To ensure that the aircrew obeyed the 3,000 feet rule, each aircraft was fitted with a barograph which registered its altitude throughout, and when it landed, umpires would inspect the instrument to ensure that the aircraft had not flown too low. If it had, the umpires would declare that the machine had been put out of action by groundfire and any information its crew had gleaned would not be passed on to its army. (A similar rule was applied many years later to Irish Air Corps aircraft during large-scale manoeuvres in 1942, also in Munster.)

The ground crew at Rathbane 'kept 'em flying'.

By late September the 'war' had ended and the aviators enjoyed some rest and recreation. This included a visit to the races at Newcastle West, where their aerial demonstration was enthusiastically received by the punters. The return to base was not without incident – en route for County Down, and into position for the sea crossing, there were a number of forced landings. The airmen landed at Newcastle in County Down and were royally entertained by the Royal County Down Golf Club, though one pilot missed his lunch by landing too far away. The following day bad weather prevented a departure, but eventually the four serviceable aircraft remaining from the original magnificent seven got away and arrived safely back at Montrose.

The long-distance return flight and the reconnaissance operations during the manoeuvres were adjudged a great success – except, perhaps, by die-hard cavalrymen who saw their traditional scouting function threatened. Mounted scouts would take forever to obtain the information so swiftly gathered by the airmen. The Irish experience was soon put to practical use: within twelve months, Number 2 Squadron was in France operating a mixed bag of reconnaissance aircraft in support of the British Expeditionary Force.

After the Munster exercises, this BE2a was used by the commander of the Rathbane squadron, Captain CAH Longcroft, for a non-stop flight of 650 miles, which won him the 1913 'Brittania Trophy'.

After the war, the battle-hardened squadron was disbanded, but it was quickly reformed at Oranmore, County Galway, to participate in the War of Independence, to the limited extent that aircraft proved effective in the guerilla warfare which ensued. The reception for Number 2 Squadron was very different from that received in 1913 when it arrived in Limerick – but times in Ireland were different too.

V

THE RFC RETURNS

BY 1917, THE ROYAL FLYING CORPS WAS A FAR MORE WARLIKE FORCE than the representatives of Number 2 Squadron in their flimsy machines which appeared in Munster in 1913. But nonetheless it was being hammered by 'The Fokker Scourge' as the intrepid German airmen, in superior aircraft, were knocking its aircraft out of the skies over the Western Front. Early in 1917 it was decided that the RFC should be expanded urgently to 106 Service Squadrons and ninety-seven Reserve Squadrons. This called for many additional training aerodromes, both in Britain and Ireland. The man chosen to survey and select suitable sites in Ireland was a Major Sholto Douglas, who had a tenuous Irish connection, as his father was curator of the National Gallery in Dublin.

Sholto Douglas had a French farm horse to thank, or blame, for his Irish mission. While taking off from his French aerodrome he collided with the horse and was injured. He spent the summer recuperating at his father's home and was then pronounced fit for light duties. As he describes in the first volume of his autobiography, *Years of Combat*, published in 1963, his new duties were a far cry from the Western Front:

> The job that I was given was that of travelling all over Ireland with the object of selecting eight sites for aerodromes on which the RFC could establish some of their training schools. There were no aerodromes at all in Ireland at that time, and the Lands Officer at Irish Command had picked out for me what he thought might be some likely sites. It was left to me to inspect them, and to make any recommendations that I saw fit. We were looking for grass fields which would give us good runs of about five to six hundred yards in any direction.
>
> Before I left for Ireland I was told that a car and a driver would be

provided for me by Irish Command, but I was thinking in terms of the air. An aeroplane would not only provide a much better means for inspecting potential sites; it would also save a lot of time. I was given permission to use a BE2c and I started off for Dublin by way of Stranraer and Larne, making the narrowest possible sea crossing over the North Channel. I landed in the Phoenix Park in Dublin and I left my aircraft outside the gates to the Viceregal Lodge, solemnly warning the two sentries on duty that they were to guard it with their lives. The troubles of only a few months before [the 1916 Rising] were fresh in my mind, and I was cautious about sabotage.

But severe though the troubles had been, I was being over-cautious, for the way in which I was treated during my tour of inspection gave me no cause at all for any alarm. Most of what are today the principal aerodromes in Éire and Northern Ireland were selected by me on that trip during the early summer of 1917, and they included such places as Collinstown, Baldonnel and Aldergrove. I also found a site for an aerodrome near Fermoy, and another one at the Curragh. I would fly over the sites that had been suggested to me, weigh up their possibilities from the air, and then go by car to make inspections on the ground.

Whenever I made any journeys by road the 'cornerboys', lounging in the villages, almost invariably scowled at me, and a few of them even threw some stones. The sight of the uniform of the British Army was becoming an increasingly unpopular one. But such is the extraordinary and at times rather touching illogicality of the Irish mind that when I flew anywhere and landed in a field near some village or other, I found that the young men would come running out onto the field cheering and shouting wild Irish cries and throwing their hats in the air, all because most of them had never seen an aeroplane before. I often wondered how many of them had thrown stones at me when I passed through the villages in my car? On the completion of this survey of Ireland I flew back to London. I made my report, and that led to airfields being established at most of the sites that I had selected, along with training schools for pilots and observers.

Construction at the sites selected by Sholto Douglas was based on a standard plan, and large permanent hangars and buildings were soon being erected. The first aerodromes to be commissioned, at the Curragh Camp, Baldonnel and Gormanston, were operational by October 1918, though the Armistice was signed a few weeks later. Apart from training,

The Royal Naval Air Service (RNAS) operated anti-submarine airships from Johnstown, County Wexford.

the principal role of the RFC in Ireland was in helping to combat the German submarine offensive. In this task it was joined by the Royal Naval Air Service (RNAS), which had set up bases at Johnstown Castle in Wexford and Malahide Castle, north of Dublin, from which airships patrolled the Irish Sea. At Berehaven, Larne and Rathmullan there were kite-balloon stations, which supplied observation balloons to naval vessels. However, many sailors felt that these balloons, flying as high as 3,000 feet, served merely to warn the U-boats of impending danger!

From 1917, the United States Navy air arm operated flying boats and kite-balloons from stations at Wexford, Cork Harbour, Bantry Bay and Lough Foyle, manned by a total of 3,000 personnel. It had been intended that the Americans would take over all anti-submarine aerial operations off the Irish coast, but before they became fully operational the Armistice had been signed.

On 1 April 1918, the RFC and the RNAS amalgamated to become the Royal Air Force. In Ireland its planes were used to encourage recruitment by dropping copies of a publication called *The Irish Soldier*, but it is doubtful whether this increased the large numbers of Irishmen already serving. In the month before the Armistice was signed, a fierce gale blew over Fermoy Aerodrome, destroying BE2s, De Havilland DH6s and 9s, and RE8s. Nature destroyed more RAF aircraft than the IRA ever would in the imminent War of Independence.

Left: Bow view of a Curtiss H-16 'Large America' flying boat.

Right: A brighter view of a USN H-16 than could be obtained in Irish skies – this location is off the sunny Hawaiian islands.

Left: Getting a large flying boat ready for flight, Queenstown (Cobh).

Right: Another H-16 takes to the water at Queenstown (Cobh) Naval Air Station for a test flight prior to its delivery to Wexford NAS.

VI

RAF versus IRA

THE COST TO BRITISH TAXPAYERS of the Army's pre-war presence in Ireland was as much as the net revenue of the country. In August 1919, when the British Forces were still in the midst of demobilisation after the Great War, the authorities reckoned that 'a normal garrison would suffice', and planned to reduce the force to that level in twelve months. The RAF at Baldonnel was reduced to a couple of depleted squadrons but before it reached this low level a 'Grand Aerial Derby' was held. It started at Tallaght and took in the Curragh and Gormanston before passing over Merrion Square (where RAF administrative headquarters were located) and on to the finish. At Tallaght, while the race was in progress, an aerobatic display was given to entertain the crowds before the competitors, racing for the finish line, appeared. Next day, there was more diversion when the airmen had a sports display at Lansdowne rugby grounds. But the fun and games would not last for much longer – the country was stirring again.

The 'fighting cock' was the emblem of Number 141 Squadron, RAF, stationed at Tallaght in 1919.

In London there was a high-level battle to save the Royal Air Force. The 'ten-year rule' – the assumption that there would be no major war for a decade – was in place, and at the end of each year this projected peaceful decade would start again. Such thinking fuelled a blazing row between Hugh Trenchard, Chief of the Air Staff, and Winston Churchill, who wanted the Army and Navy each to have their own airforces. But when Churchill became Secretary of State for the Commonwealth Office in 1921, he supported Trenchard's plan, which was to 'police the Empire by air power alone', using the first-line squadrons of an independent RAF to keep the peace in the Near and Middle East and in India. Trenchard had to rebuild an airforce from the remnants of its former great strength and he was severely restricted by a limit of £15 million per year for the next five years. It was against this penny-pinching background that the RAF operated in Ireland.

The Hucks starter, a device mounted on a Ford model T, which would use the engine to spin a shaft that engaged in an aircraft's propellor boss and start the aero-engine without the sweating ground staff.

The Irish Republican Army (IRA) had raided Collinstown aerodrome and made off with seventy-two rifles, complete with bayonets and ammunition, and the War of Independence had begun. Now, far from being decommissioned, the RAF aerodromes were strengthened with

'I was very struck with the turnout of the three Flights of Number 100 Squadron in formation flying, the general smartness of their appearance, and the fact that they were back in the hangars within an hour. In addition, without being in a position to criticise, it struck me that the performance was an extremely good one in view of the fact that a 30mph wind was blowing, with rain and heavy gusts.' General Macready, General Officer Commanding , The Crown Forces in Ireland. 6 December 1921.

barbed wire entanglements, watchdogs and, taking a lesson from the ancient Romans, even geese as an early warning system! Aircrews carried rifles for protection if forced to land in 'inhospitable' terrain. Number 100 squadron was reformed at Baldonnel at the beginning of 1920. It was commanded by Squadron Leader Frederick Sowrey, who had proved his mettle in 1916 by downing a Zeppelin over Essex. Number 2 Squadron was reformed (one month after it had been disbanded) at Oranmore, near Galway, from where detachments were sent to Fermoy and Castlebar.

Baldonnel, 4 February 1922. The CO of Number 100 Squadron, RAF, Squadron Leader F Sowrey (on the right), with his wife, prior to leading the final departure from Baldonnel. Sowrey flew over Croke Park on the original 'Bloody Sunday', though his flight was unconnected with the events that shortly followed.

In the north, a flight from Number 4 Squadron in Britain was detached to Aldergrove. These units totalled seven flights of three aircraft each and were generally equipped with the formidable Bristol FB2 Fighter, commonly known as the Brisfit. This type was a well proven two-seater, with a forward-firing Vickers machine-gun operated by the pilot and a free-swinging Lewis gun in the observer's cockpit behind. It could carry a dozen small bombs or a pair of larger ones, totalling one hundredweight in either configuration. It appeared ideal for the job in hand, but was the task suitable for *any* type of aircraft? It would be fifty more years before the counter-insurgency helicopter gunships used in Vietnam were developed. As the war intensified and Army patrols were increasingly menaced by IRA guerrilla groups, known as 'flying columns',

'We'll all swing together!' Three-man teams link up to swing Bristol Fighter engines into action, normally the job of a Hucks starter.

the question of using air support was examined closely. On the ground, the Royal Irish Constabulary (RIC) was reinforced by the notorious 'Black-and-Tans' and the more professional and deadly RIC Auxiliary Division, the 'Auxies'.

The regular Army could no longer use the normal postal services and called upon the RAF to provide an airmail service to the various aerodromes or a suitable field close to military outposts. Where a landing could not be attempted, the mail was dropped – not always accurately, nor to the designated recipients! The *Irish Independent* reported:

> A band of Sinn Féiners, in the guise of British soldiers, have captured official mails taken by aeroplane to the military at Bantry. It had been customary for aircraft to drop the bag onto a wide circle marked on the ground where waiting soldiers took charge of it. Near to the military quarters in the Bantry district on Sunday afternoon last, 12th, the Sinn Féiners copied the device with success. They made a large white circle which the aeroplane pilot mistook for the destination of his mails. He dropped the bag which the Sinn Féiners seized.

The garrison at Waterford Castle was similarly unlucky when an aircraft dropped its mailbag into the sea! Later, in the same city, another aircraft was attempting to drop mail into the barracks square but approached too low and hit rooftops opposite, resulting in both occupants being seriously injured.

At Bantry, where an unsuitable landing strip had been prepared, an aircraft was demolished on take-off, though the pilot and his passenger, a senior staff officer, escaped unhurt. Aircraft were frequently used to ferry senior Army officers, one of whom reported that: 'To go from Dublin to Cork, one may fly, one may go by destroyer and be met by an escort at the docks, or one may go very slowly by armed train.' This officer, Brigadier General Crozier who commanded the Auxiliaries, always travelled by air, having once been blown out of his motorcar by a landmine. When using 'air taxis', peppery bespurred cavalry colonels didn't always approve of RAF treatment. The following signal from RAF Wing Headquarters warned:

> It is to be understood that when Army or other personnel are carried as passengers in Service Aircraft they are to come under the orders of the officer making the arrangement for the flight or the pilot of the machine. A recent case in which an Army officer objected to Royal Air Force mechanics helping him into a machine is one which, should it occur again, is to be reported immediately to Wing Headquarters. The crews of aircraft, when giving assistance like that mentioned above, are to do so with every courtesy but are responsible that no damage is done to their

> machines by officers or others, not belonging to the Royal Air Force, who
> are not accustomed to aeroplanes and are inclined to be careless,
> especially in getting into or out of machines.

The increased workload as maintenance escalated reduced the
availability of aircraft; at one stage only three aircraft could be flown.
Winston Churchill looked for answers from Trenchard and the latter
responded that the aircraft were old and had not been properly
maintained in the past; that the canvas hangars, still in use at various
bases, could not stand up to the severe Irish climate; that accurate
logbooks (the *CV* of each aircraft) were not being properly kept; and
finally, that the organisation of stores was chaotic and suffered from the
non-receipt of spares from Britain. Trenchard added that there were no
proper facilities in Ireland for servicing and repair work and that,
currently, twenty aircraft there were in need of a complete overhaul.
Trenchard also rejected the request of General McCready, who
commanded the Army in Ireland, that such serviceable aircraft as there
were should be armed and used offensively. Trenchard judged that these
should not be employed until 'a definite state of war is declared to exist in
Ireland'. Apparently he was reluctant to admit that a 'war' was in progress.

At Churchill's prompting, the Secretary for Air visited Ireland and
his brilliant suggestion was that the Wicklow Mountains should be
cleared of all their inhabitants to facilitate bombing practice there!
Providing air reinforcements was difficult – crossing the Irish Sea was still
an adventure and could only be undertaken by very experienced pilots; a
number of ferry pilots had already been lost at sea. Action was needed,
and though Trenchard's squadrons had a major job in hand 'pacifying' the
natives in India, Iraq and elsewhere, he had already grudgingly sent over
one additional flight.

In County Cork, where the War of Independence was being most
vigorously pursued by the IRA, Fermoy aerodrome was virtually useless
because it had no repair facilities. Trenchard maintained that the
solution lay, not in providing extra aircraft, but in improving maintenance
facilities for the ones that were already there. He added that because of
overseas commitments only trainee squadrons were available. Gradually,
air power was increased and active operations against the IRA
commenced. By the middle of March 1921, Number 100 Squadron was

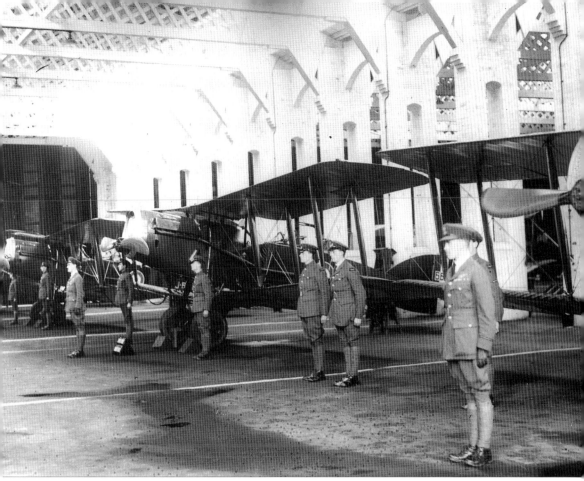

Ready for VIP inspection – Number 100 Squadron at Baldonnel during its last months of service in Ireland.

cooperating with the ground forces engaged in 'sweeps' to round up the 'flying columns', by dropping messages confirming their whereabouts. Engine troubles still persisted, and many years later Group Captain Frederick Sowrey, who had commanded Number 100 Squadron, told this writer of his interview with Trenchard regarding spares. He got very short shrift – the great man's loud voice was so intimidating that he went by the nickname of 'Boom'!

Frequent forced landings occurred where the IRA was active. A plane carrying mail from Fermoy to Tralee force landed at Kanturk, and was surrounded by soldiers, who were then attacked by guerrillas, with both sides suffering casualties. In the same month, August 1920, another force-landed aircraft was burnt out while its crew sought assistance. In County Clare, military guarding a crashed aircraft were engaged, but no casualties were reported. In the townland of Cooladoody, County Waterford, in a similar attack, the guerrilla fighters managed to get to the aircraft and set it on fire.

Men of the West: A typical flying column drawn from the Irish Republican Army in Connemara in 1921.

A souvenir of these incidents can be seen in the National Museum – the four-bladed propeller of a Bristol Fighter, serial 1487, which, according to the war diary of Number 2 Squadron,

> … was forced to land at Kilfinnane, County Limerick, owing to engine trouble, on an airmail flight between Fermoy and Oranmore. It was decided that the pilot would remain by the machine while the observer proceeded into Kilfinnane village to obtain a guard for the machine. During his absence, a party of about twenty-five rebels, armed with service rifles, attacked the machine and called on the pilot to surrender. They set fire to his aeroplane and made off with him. When the observer returned with a police guard he discovered the machine completely burnt out. Later, messages were received from the local parish priest stating that he could vouch for the safe return of the pilot who, on his release, confirmed that he had not been ill-treated in any way.

Before he was released, his fellow flyers and troops on the ground had been out in strength looking for him and his captors, but none of the rebels, the East Limerick 'flying column', was apprehended. When RAF aircraft were destroyed, as in this case, the British authorities made a malicious damage claim on the relevant local authority – and the poor ratepayers had to foot the bill. The air war, such as it was, and other hostilities ceased when a truce was called on 11 July 1921, and an uneasy peace reigned.

'The descent of Mr Livingston on the Coast of Baldoyle, County of Dublin, on Thursday the 27th of June 1822, on which day he ascended from Porto Bello Barracks for the benefit of the suffering Poor of the South & South West of Ireland.'

For the film epic The Blue Max, replicas of Great War aeroplanes were built.
This is the deadly Fokker DI triplane.

Left: The first of a fleet Aer Lingus Airbus Industries A330s, which now numbers twelve. type has the best operating costs in its class, while offering passengers a quiet and spacious cabin. Typical seating 335 passengers Airbus Industries stand first and economy layer the A330 is aimed at medium to long-haul routes. It is claimed to the most efficient aircr ever designed.

Right: No more popular name ever took to the Irish sky than that of 'the product' on this Starduster.

Left: Flying machines, however modern, com in some odd configurations, like this Thruster T300.

Right: A formation of SIAI Marchetti SF 260 WE 'Warriors'.

Below: one of the type in more belligerent mood, as it unleashes a salvo of 68mm SNEB rockets at a target at sea, at the air firing range at Gormanston, County Meath.

Above: The five Eurocopter Dauphins of Number 3 Support Wing.

Left: 'Gazelle in the Moonlight': A light training helicopter at the end of a day's work.

The eagle flies again: When Aer Lingus was planning its Golden Jubilee in 1986, it refurbished an Irish DH Dragon to replicate its first aircraft.

A pair of Aer Arann 'Islanders' overfly the prehistoric Dun Aengus on Inishmore. Perched on 300ft cliffs, the fort is judged to be the most magnificent ancient monument in Europe. Apart from serving the Aran Islands, Aer Arann is also an efficient internal airline.

Planes on parade, Baldonnel, April 1979. The Air Corps was heavily involved in events surrounding the Papal visit – Fouga 'Magisters' escorted the incoming Papal Aer Lingus flight, helicopters were used for casualty evacuation and VIP transportation, and armed 'Warriors' constantly patrolled restricted airspace above Papal sites and temporary airways.

'The government jet' is the public's name for aircraft of the Air Corps's Ministerial Air Transport Squadron. Here are two: Numbers 249, a Gulfstream GIII, and 238, a Bae 125–700.

An Air Corps CASA CN 235–100MPA 'Persuader' maritime patrol aircraft overflies the Naval Service helicopter carrier LE Eithne. More aircraft and vessels are needed to fulfil EU commitments.

'Let there be light!' Illuminated by a parachute flare, the giant Sikorsky flying boat Excambrian escapes the reluctant waters of Foynes flare-path in this faithful reconstruction by Vincent Killowry, Ennis. Where possible, a nighttime take-off was preferred because this would permit a daylight landing (a more difficult operation) on the American side, and would also provide the aircraft with a cloak of darkness against the marauding long-range Luftwaffe Condors.

Owned, flown and here photographed by George Flood, Dublin, this immaculate DH82a Tiger Moth is any pilot's dream!

AN POST MARKS THE MILESTONES OF FLIGHT

Ireland's current definitive stamp series beautifully depicts the country's wide and varied bird life. By a happy coincidence, recent sets of commemorative stamps depict man's endeavours to emulate his feathered friends in Irish skies.

The top two stamps honoured Richard Crosbie in 1785,
while the fiftieth anniversary of the first east–west Atlantic crossing was marked by
two denominations, one brown (seen here) and the other blue.

At either end of the next row, the twenty-fifth anniversary of the founding of Aer Lingus is featured,
while between these there are two attractive pictorial views released to make up the transatlantic postal rate.
In addition to this pair, three other famous landmarks connected with early Celtic Christianity were issued,
all featuring the Angel Victor (the messenger of God to St Patrick) carrying Vox Hibernia (the voice of Ireland).

The third row begins with the epic flight of Alcock and Brown in 1919, and the adjacent stamp shows
a pick-a-back composite flying boat at Foynes. The remaining two stamps in this row commemorate
Aer Lingus's Golden Jubilee by showing how it developed from the DH 'Dragon' of 1936 to the Boeing 747 of 1986.

The fourth row honours three outstanding airmen and one airwoman, Lady Mary Heath from County Limerick.

The fifth row is given over to notable aircraft, leading with the redoubtable Douglas DC3, which, after World War II,
was the aerial workhorse that was crucial in the development of Aer Lingus.

The last two sets were designed by Vincent Killowry, and are reproduced here by courtesy of An Post's Philatelic Bureau.

A further set of stamps will be issued late this year, featuring Irish Air Corps aircraft, both past and present.

VII

'THE BIG FELLA'

WITH THE CALLING OF A TRUCE IN JULY 1921, the reign of terror by the Crown forces ceased and the plain people of Ireland hoped for a lasting peace. The ceasefire was followed by a series of meetings in London between Éamon de Valera and the British prime minister David Lloyd-George. A full conference was arranged, but de Valera startled his colleagues by announcing that he would not participate and he recommended that Michael Collins should go in his stead. Collins, who had led the armed struggle and particularly the intelligence battle, had good reasons for not wishing to do so. He said: 'It is an unheard-of thing that the soldier who fought the enemy in the field should be selected to carry out negotiations.' In the end he went and was to say later: 'I went to London with a heavy heart, in the spirit of the soldier sent to do an unpleasant task by his superior officer.'

Collins had evaded capture for years, but his IRA comrades thought that he would be vulnerable if there was an impasse at the conference. Emmet Dalton, a close ally of Collins, devised a scheme for a quick getaway by air. The plan was initiated when a jaunty red-haired Volunteer, a former pilot of the wartime RAF, presented himself before the

General Michael Collins with Major General Emmet Dalton, who devised the get-away aircraft plan, and in whose arms Collins was to die later at Beal na Bláth.

president of the Irish Self-determination League, Art O'Brien, in his office by the Strand in London. The pilot introduced himself as Charlie Russell and requested funds to buy an aeroplane. Though O'Brien was accustomed to providing money for various projects, including the purchase of arms, he demurred somewhat, but Russell's credentials persuaded him to part with a cheque for £2,600. Russell had already selected the aeroplane at the Aircraft Disposal Company (ADC) at Croydon. In addition to large numbers of redundant wartime aircraft, this company had acquired the stock of the Martinsyde Aircraft Manufacturing Company, which had gone into liquidation. This included five Type A Mk.II passenger aircraft, which could accommodate four people in an enclosed cabin, behind which was the pilot's open cockpit. The aircraft had a duration of five hours at a cruising speed of 100mph. Obviously Russell didn't disclose the true nature of his purchase to the ADC, but posed as a representative of the Canadian Forestry Department – he could affect a plausible accent, having spent some time in Canada.

The Martinsyde MkIIA, its assembly completed at Baldonnel in October 1922. The pilot's cockpit is on the left; centre is the glazed hatch that the four passengers reached by means of an aluminium ladder.

As the talks at Downing Street proceeded, Russell needed to delay delivery of his purchase for as long as possible. He embarked on a lengthy series of test flights and, after each one, managed to convince the vendor that there were problems – port wings rigged too high, or too low; suspected tail flutter; dud oil pressure. He proved to be a thoroughly difficult customer, but always ensured that the aircraft would be ready if needed. There had been, incidentally, another scenario in the original plan – the Martinsyde was supplied complete with floats to convert it into a seaplane, and in this configuration it could have been based on Southampton Water for a short flight to France. In the event, the conspirators decided on a flight plan which would take Collins and three of his aides southwest to Bristol and, skirting the Welsh coast, across the sea to Rosslare. From here, Russell would follow the railway line (in those days railway lines were *the* great aid to navigation) north to Dublin and a landing on Leopardstown Racecourse. A reception party commanded by Jack McSweeney, another ex-RAF pilot, would be deployed.

Charlie Russell signalled Emmet Dalton:

> McSweeney will be able to give all instructions for the landing, but it must be clearly understood that the two men who are to catch the wings and rear struts must make sure not to catch the *edges* of the wings. If it so happens that they misjudge the distance, it must be impressed upon them to fall flat on the ground and let the aeroplane pass over!

Russell's instructions continued:

> In the event of our having to cross at night, it would be necessary for your people to mark out an L-shaped figure upon the landing ground; this should be done by means of four petrol-tin fires or flares at intervals of sixty yards. One flare should be placed in order to indicate the direction of the wind with the remaining three placed accordingly. The cloth signal to be used in daylight is also a wind signal; the triangular piece should be placed indicating the direction in which the wind is blowing. I will be able to notify you fully of the time at which departure will take place, and the journey is calculated to take between three and four hours.

At the end of the year, the Anglo-Irish Treaty was signed, making the escape plan redundant. Under the terms of the Treaty, Ireland was partitioned, with the northern six counties remaining part of the UK. The Irish delegation returned by sea, to the jubilation of many, but to the

The Martinsyde, bearing the name The Big Fella on a scroll on its cowling, is seen at Baldonnel in February 1923.

reproaches of dissenters who wanted nothing less than an independent republic, as indeed did the delegates had this goal been attainable.

Russell could now come clean and arranged that his aircraft be crated and shipped to Dublin – perhaps the vendors thought that he was reluctant to fly it over after his long litany of complaints! In June 1922 the aircraft duly arrived at Baldonnel from Dublin Docks. Russell, who had been made responsible for organising civil aviation,

Charlie Russell as a Major General in charge of the Railway Protection, Repair and Maintenance Corps in 1923. He returned to the Army Air Service later as OC.

recorded its name, *The Big Fella*, in its log book. This was Collins's nickname, a name which had arisen from his dynamic, pushing ways. He liked to throw his weight around and get things done. One of his biographers, Frank O'Connor, noted: 'The story of his brief life is the story of how he turned the scornful nickname into one of awe and affection.'

Collins had a keen appreciation of the value of aeroplanes, both in war and peace and, before the truce, had already planned to capture one of the RAF's planes which he would use, in his own words, to 'disturb the Black-and-Tans in their strongholds'. *The Big Fella* was not the first aircraft to arrive for the newly-formed Irish Air Service, part of the new National Army. However, by October 1922 it had been assembled and finished in silver dope, with the national colours applied to its fuselage and tail and the legend *The Big Fella* to its nose. Ironically, when the craft was ready for air testing, the pilot, one Captain Tom Arnott, was uncovered as an

'The Big Fella', General Michael Collins (1890-1922) as Commander-in-Chief of the National Army.

ex-Auxiliary! He was unceremoniously escorted at pistol-point to the mailboat. Another irony: Art O'Brien, who had funded the purchase of the Martinsyde, opposed the Treaty, and refused to hand over the assets of the Self-determination League to the Provisional Government of the new state. Collins, in his capacity as Minister of Finance (he was also Commander-in-Chief of the new National Army), was forced to take a High Court action in London to retrieve the disputed funds.

The split in the IRA had not yet developed into civil war. Nevertheless, the Air Service considered trading in the Martinsyde for more warlike planes. When two Bristol Fighters were taken over from the RAF in early July, however, the British noted that no armaments were requested. Thereafter, the Martinsyde was not highly utilised, being flown in the early 1920s, mainly by the then commanding officer, Commandant Tom Maloney. He always called for the same passenger – a well-nourished private who provided adequate ballast for the five-seater! Maloney believed he could have used it, with its passenger compartment given over to a large fuel tank, in an attempt on the east-to-west Atlantic crossing which was on everybody's mind at the time. Indeed, Commandant James Fitzmaurice, who succeeded Maloney after the latter's death in a crash, offered the Department of Defence £1,000 for the aircraft with the same plan in mind, but the offer was rejected. The Martinsyde was flown occasionally before being relegated to the status of an instructional airframe in October 1927.

With the change of government in 1932, Collins's nickname was painted out and the name *City of Dublin* was substituted on the aircraft. Further downsizing of the Collins legend took place in the 1950s when a recruiting drive for the Air Corps included a short history to be broadcast

Wingless and engineless, the Martinsyde languishes in a hangar, its original name having been changed to City of Dublin – in Gaelic on the starboard cowling and in English on the other side. In 1935, this historic aircraft was completely demolished.

on Radió Éireann. A senior air officer prepared the script and submitted it to the appropriate government department for clearance. He was asked to alter just one word: could he perhaps change the original name of the Martinsyde to *The Long Fella*, which was a nickname sometimes applied to Éamon de Valera? The officer refused, and the broadcast went out as per his script. *The Big Fella*, as everyone still called it, languished forgotten in a corner of a hangar until 1935 when space considerations dictated that it should be scrapped. When representatives of the National Museum of Ireland later investigated this unique relic, nothing remained but its log book and a few souvenirs, notably the propeller boss which, suitably inscribed, has an honoured place in the Officers' Mess at Baldonnel. In these more enlightened days of aircraft preservation, what a prize exhibit the State's first aircraft would be!

VIII

AN IRISH AIR FORCE?

EARLY IN APRIL 1922, the first units of a small regular National Army, as envisaged by the Provisional Government, were forming in Beggar's Bush Barracks, Dublin. From there a detachment marched out to an unceremonious takeover of Baldonnel aerodrome from the Royal Air Force. Under the terms of the Treaty, British forces were to withdraw from the Free State. The departing RAF left little behind them, their final act being to drive trucks over surplus equipment. In Beggar's Bush Barracks Jack McSweeney and Charlie Russell, the two young pilots who

October 1922: The first enlisted man, Sergeant Johnnie Curran, sits in the front cockpit, while Lieutenant Bill Delamere waits for the prop to be swung.

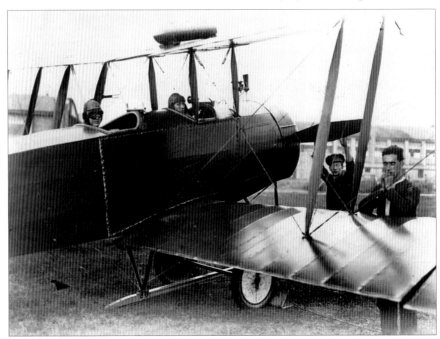

had been involved in the *Big Fella* plan, now began to establish an air arm as part of the Army. They named it the Irish Air Service, and set about interviewing recruits. The first recruit was an ex-RAF mechanic, Johnny Curran, who was straightaway given the rank of Sergeant. The three began to assess facilities and organised workshops with the help of the Baldonnel garrison.

Once word got around about the establishment of a new air force, other ex-RAF pilots joined, in dribs and drabs. A number of the recruits had fought over the Western Front and elsewhere, and one had held a short service commission in the post-war RAF. Most of them had spent the three years since demobilisation finding their feet in civilian life, a dreary existence after the excitement of war. The prospect of flying again, and being paid for it, was a godsend. They still retained memories of their 'death or glory' days, when they sang:

> So stand by your glasses steady,
> The world is a world of lies.
> Here's a toast to the dead already,
> And here's to the next man who dies!

Eventually the pilots numbered thirteen, but Tom Arnott was uncovered as an ex-Auxiliary and swiftly booted out. Flying skills varied, as some

The first fighting machine, a 'Brisfit'. The national colours have been applied, although the RAF roundels are still on the lower wings.

pilots had grown rusty since the war. In his first month Jack MacSweeney, now a Commandant-General commanding the Service, made a bad forced landing near Naas, County Kildare, slightly injuring himself and his observer – all because of a lack of practice on the Bristol Fighter type. Charlie Russell, now Director of Civil Aviation, attempted to loop-the-loop in another Bristol Fighter and almost fell out of its cockpit at the top of the loop! Later, he had a more serious mishap. One of the crated Avro trainers had been erected by an ex-RAF sergeant major, who had accidentally reversed the controls; it later transpired that he was a regimental rather than a technical type. Russell took up the Avro with a private who had helped in rigging it, only to hit the side of a hangar, causing both severe injuries, which later resulted in Russell losing an eye.

A considerable number of Irishmen had joined the Royal Flying Corps (RFC), and some of them had become 'air aces' by shooting down five or more enemy aircraft. One such flyer was George McElroy from Donnybrook; he had served in the trenches before returning home to recuperate from being badly gassed. He transferred to the RFC at the beginning of 1917 and, after training, was posted to a fighter squadron in France, where he scored his first victory at the end of the year. During the next seven months his score shot up to an astonishing forty-one victories and he was awarded the Distinguished Flying Cross and Bar. He was a comrade of another outstanding pilot, Mick Mannock, and they tended to admonish each other for being too aggressive and for going too low in pursuit of German aircraft. However, Mannock was to die doing just that, when his score had reached sixty-one, and McElroy was lost five days later, also during a low-level pursuit.

The newly-established Irish Air Service attracted a number of these flying aces. One of the recruits was Oscar Heron from Armagh. Heron had served for the final six months of the War, and in that short time had shot down a total of ten German aircraft, for which he had received the Distinguished Flying Cross and other awards. Some of his colleagues in the Service had yet to fire a shot in anger, but very soon they would be in action again, sadly against their own countrymen. The bitter Civil War, which resulted from the division of the people into two sides – pro-treaty and anti-treaty – was about to begin. The ex-pilots brought with them their anglicised accents and the hard-drinking habits acquired during the war. In contrast, their observers/gunners, though by no means teetotal, were from a different background – IRA men who had taken the Treaty side.

Seizing the photo opportunity, the ground staff pose in front of Avro Serial AIII, seen on page 56, which flew on in its peacetime silver finish for a further nine years.

The essential technical trades vacancies were filled mainly by motor mechanics, along with a couple of technicians from the British air arms. One of the latter described the chaotic scene which greeted him at Baldonnel when he arrived:

> The place abounded with armed troops, and weapons were being discharged indiscriminately … In one of the billets, a chap was using his rifle to hammer out his initials on a partition wall, blissfully unaware that his bullets were travelling through the adjacent billets and scaring the hell out of their occupants!

The situation gradually improved. Four companies of Air Service Infantry were formed to protect the aerodrome from the Irregulars, the anti-treaty side, who had conceived a plan to steal an aircraft and bomb the seat of government at Leinster House. In later years the 'bomb aimer' of the plot described this plan as high farce. He enquired about it from the pilot, known as 'the Deacon' (he had advanced to that stage of holy orders before espousing the anti-treaty cause). The Deacon confided that after the attack he proposed to land on Merrion Strand, but admitted that he had not checked out the times of the tides. Not surprisingly, when the plan to attack the aerodrome was finally cancelled, the 'bomb aimer' was greatly relieved, particularly as other aerial targets had been proposed. These were army barracks, where fires would be lit by subversives to guide the rebel airmen. Though the raid to appropriate the aircraft, as many arms and as much ammunition as possible was called off, casualties still resulted. Three members of the aerodrome garrison who had deserted to join the Irregular raiding party were captured, court-martialled, and shot.

Before the Treaty, RAF aircraft had made many forced landings and one of its aircraft had had more than its share of bad luck. This was Bristol Fighter serial number E2411, one of two donated to the new Service by the RAF's Irish Flight which was still occupying Collinstown. The Brisfit was well used to Irish skies and indeed Irish fields! While on the strength of 141 Squadron, it had force landed near Lucan. Later, when based at Castlebar, it suffered a crash landing. Repaired, it went to Fermoy where, its engine having cut out on take-off, it collided with a cow! It was taken over by Charlie Russell who, to underline his position as Director of Civil Aviation, dressed in civilian clothes topped off with a bowler hat. Perhaps the British were glad to get rid of this unlucky aeroplane? On reaching Baldonnel it was given a new name, 'Number One', and it had much better luck with its new owners, surviving until 1924 despite heavy utilisation.

The Civil War, which had been simmering for weeks, finally broke out as Russell was taking delivery of the Brisfit. He was ordered to support the National Army surrounding a detachment of Tipperary Irregulars at Blessington. Without the normal free-swinging Lewis gun in his cockpit, his observer, Captain Bill Stapleton, had to use the unwieldy infantry version. Over the scene of the fighting, Stapleton balanced perilously on the cockpit edge, virtually firing from the hip. The action at Blessington lasted for just six days as the troops tightened their noose around the town and trained field guns on it. Brisfit Number 1 contributed to the fighting, which ended when the besieged men slipped through the encirclement, leaving a dedicated fifteen-man rearguard to cover them. By now, further purchases from the Aircraft Disposals Company were arriving at Baldonnel: more Bristol Fighters, Avro 504K trainers, Martinsyde F4 Buzzards (a first-rate single-seater fighter, which had been just too late for the Great War) and a lone SE5a, which was the Spitfire of that war. Brisfit No.1 now moved down to the Fair Green in Limerick City, from where the Irregulars were retreating to regroup and hold the line of what they called the 'Munster Republic'.

Though Bristol Fighter Number One went into action at Blessington with a hand-held infantry gun, eventually a proper aerial Lewis was fitted mounted on the usual Scaarf Ring, as seen here.

From their temporary base the two airmen reconnoitred enemy movements, which they reported back to General Collins (Captain Stapleton had been a member of Collins's carefully-picked 'Squad'). On one mission, incorrect fuel caused engine trouble on take-off from the green, and on making a hasty return there the aircraft was put out of action. While repairs were under way Russell returned to Baldonnel. It was here that he received General Collins's last signal (Collins was killed a few days later), notifying him that Fermoy Aerodrome was suitable for landing and specifying the areas that he should urgently reconnoitre. Earlier, a company of the new Army had taken over the aerodrome. Its commander recalled: 'When the Union Jack came down and the Tricolour was hoisted there, the crowd gave us a round of applause. That stirred up a small bit of pride in us too, that we had raised our own flag up in Fermoy aerodrome.' However, soon afterwards, he and his company decided to throw in their lot with the Irregulars!

It was not until October that three aircraft were despatched from Baldonnel to Fermoy to investigate the lie of the land. What they found was not encouraging. During the six months which had elapsed since the RAF had departed, local farmers had used the unguarded aerodrome for tillage and had also made off with all the corrugated sheeting from its three hangars – all that remained were their non-portable steel skeletons and huge doors. Even the lighting plant had disappeared! Troops and aircraft scoured the countryside, and found every single sheet of the distinctive black corrugated cladding, which was soon back in position. The hangars were reconstructed by a contractor and a locally-recruited works and service company. Once more, Fermoy Aerodrome was fully operational.

From there, air operations commenced; these were mainly forward reconnaissance of the countryside, which the Army had divided into 'battalion areas'. When the columns of the Irregulars were located, the airmen engaged them with light bombs delivered over the side by the observer, who also brought his Lewis gun into action. As some of these rear-mounted guns had no collector bags to take the red-hot spent cases, these would spew onto the neck of the unfortunate pilot, thus causing all kinds of unexpected aerobatics! Less lethal were leaflets dropped from the planes offering a free pardon to those who surrendered their arms by a certain date in December. *The Irish Times*, describing a typical air action over an ambush, reported thus:

> When the lorries were passing, intensive fire was opened on them. One soldier was killed. The troops replied with fire but realised that they were outnumbered and that their attackers were in excellent defensive positions. Reinforcements were sent for and soon an aeroplane arrived and its pilot descended to a few hundred feet, and when just over the ambush party, nose-dived in their direction. Bombs were dropped from the aeroplane and the men were scampering in ones and twos towards

A mixed bag at Fermoy, 1923. A Martinsyde Scout, a Bristol Fighter and a brace of DeHavilland 'Nines', with pilots and observers in the foreground.

the wood, seeking any shelter it might afford. The plane however circled the wood and raked it with machine-gun fire. The dramatic employment of air force against attacks on national troops has made a great impression here in Cork.

An entry in a pilot's logbook was more laconic when referring to another action: 'Fired on by Irregulars. Returned fire and dropped two bombs; held enemy in houses until dark; troops approached and attacked; one killed, two wounded.'

There was one classic example of an Air Service 'own goal'. In response to a signal calling for urgent air action against a party of Irregulars who were mining a bridge, a Bristol Fighter was quickly started up and, with no time to load the bomb racks, a twenty-pounder was handed up to the observer. Over the target, and over-anxious to get rid of the lethal load on his lap, the observer despatched the bomb prematurely at a very low altitude. It exploded, missing the enemy but demolishing a section of the bridge and peppering the aircraft with its shrapnel. The Brisfit staggered back to Fermoy with its fabric in ribbons and the observer's face heavily punctured with fragments from his windscreen – this did not worry him half as much as the impending wrath of his pilot!

The extremities of Kerry, where very wild terrain concealed strong Irregular forces, were not within range of the airbase, so a Brisfit was despatched to a landing strip beside Tralee Barracks until Fermoy was reinforced with De Havilland 9 bombers. The 'Nines', introduced in pairs in 1923, could cover the whole southwest area. One crashed on returning from a patrol and, though its pilot suffered only minor injuries, the observer was killed instantly – the first and only Irish Air Service flier ever to be lost on active service.

For over a year, the planes constantly patrolled the railway lines, which were frequently breached by the Irregulars, and also cooperated with the Army in its 'sweeps' to round the guerrillas up. Despite uncertain engines and treacherous terrain, no further casualties were suffered. By the spring of 1923 the Irregular forces, despite fighting fiercely, had been worn down by sheer weight of numbers, and the Civil War ground to a halt. A pilot wrote *'finis'* in his logbook to a period which had set brother against brother and left a bitter legacy that was to plague the country for many years. This pilot noted: '14 April 1924; 05.30 hours; Bristol Fighter; Fermoy to Baldonnel; Fermoy closed; visibility very good.'

At Baldonnel, the Irish Air Service now looked to the future with confidence – aviation held exciting peacetime prospects. But there was a hiccup: there was unrest in the National Army in 1924 among a group of disgruntled officers who had been in the forefront of the War of Independence. They felt that they were being sidelined into backwater appointments where they would have little say in Army policy or training. These hardened warriors felt that people who had been less active in the struggle were being given preference, and they also complained about the rapid demobilisation of the large army which had fought in the Civil War. An ultimatum was served on the government and a military coup d'état seemed likely. However, a distinguished senior Army officer acted as go-between and the result was a major walk-out by both senior and junior officers and the 'mutiny without malice' fizzled out. 'The absconders', as they were known, included thirteen officers, mainly non-flying types, from the Air Service.

In that unsettled year of 1924 the Service was renamed the Army Air Corps, and young officers from other Army units were inducted for pilot training. Two years later the Corps enlisted the first ever Army cadet class. The sizeable number of pupils created high utilisation of the half-dozen basic training aircraft. The Avro 504Ks, unsophisticated but unforgettable, were powered by a rotary engine which pushed the aircraft along at 95mph. One of the pupils recollected:

Everything about flying was very primitive. There were no parachutes, no radio, and the aircraft had no flaps or brakes. There was no safety harness, simply a belt fastened round the waist. Communication between instructor and pupil was by means of a speaking tube connected to the flying helmets, and at times the helmet for one pupil was left *in situ* for the next one, irrespective of the size of either pupil's head! As planes were scarce, fast turn-arounds were the order of the day.

FIRST AIR MAIL
26TH AUG 1929.
GALWAY-LONDON

The first airmail service between Ireland and the UK was flown between Oranmore Aerodrome outside Galway to London's Croydon Aerodrome on 26 August 1929. The moving spirit behind this trial was Colonel Charlie Russell who, with 'Mutt' Summers, the chief test pilot of Vickers Aviation, collected American mail which had arrived in Galway on the German liner Karlsruhe.
Seen here: the Postmaster is handing up one of the seven bags of US mail to Russell at 7.30am. The Vixen touched down at Baldonnel at 8am and was en route to Croydon, via Chester, at 8.30, where it arrived at 11.37. The aircraft turned around and left Croydon at 2pm, arriving via the two outward stops at Oranmore at 7pm. Shortly afterwards a banquet was held at the Hotel Royal in Galway to celebrate the flight. Post Office officials reckoned that twenty-four hours had been saved in delivering the incoming US mail.
Insert: the Post Office cachet.

The Officers' Mess had a mixed membership – the instructors were wartime veterans who had a somewhat British 'old boy' attitude, while their pupils were either young National Army officers who were ex-IRA men, or fresh-faced cadets. The latter saw themselves as the new wave in aviation and regarded both instructors and Army students as positively ancient! Nevertheless, the unifying spirit of flying developed a very distinctive *esprit de corps*. In 1929 a demobilisation process caused many temporary commissions to be rescinded.

Above left: About two years after the first airmail service to the UK, Colonel Russell was again a mail pilot when a trial delivery was made between Galway and Berlin. The mail was first flown to Baldonnel and transferred to a KLM airliner piloted by Russell via Croydon and Rotterdam to Berlin. On this occasion the letters had the cachet shown on the photograph of the aircraft about to leave Baldonnel.
Insert: the Post Office cachet.

Later some officers, thinking perhaps of promotion prospects, transferred to other, less dangerous, units. Inevitably, in the business of flying, death also took its toll.

After the Civil War, the GOC and Inspector General of the Army, the controversial Eoin O'Duffy, proposed a peacetime army of some 28,000 men, but dismissed the concept of military aviation. He declared that its practical utility during the Civil War had been negligible (he was re-echoing opinions on the RAF's contribution during the War of Independence). O'Duffy considered that: 'The smallest aerial unit of 150 men would be sufficient to keep progressive thought stimulated, give our troops a knowledge of the value of aerial co-operation, train a small number of infantry as pilots, and keep an eye on developments elsewhere.' His proposals regarding the strength of the national army came to nothing – by 1934 it had diminished to about 6,000 all ranks, with largely notional reserves – its lowest level ever. Neglect had also brought the Air Corps down almost to O'Duffy's suggested strength, a mere 220 of all ranks. The Civil Service had virtual control of the Army. It held the purse strings and begrudged every penny spent on it.

Though a handful of modern training aircraft were acquired in the 1920s, the Air Corps was stagnating. By the end of that decade its non-flying commander, Major T Liston, reported to the Department of Defence that his unit, '... as presently organised, is a worthless organisation, costing £100,000 to run annually, with junk equipment and insufficient mechanics'.

The Air Corps' suggestion that it should become the nucleus of a national civil aviation service, and be employed not only in support of the Army but also on duties such as fisheries protection, received no official support. As World War II would soon show, the overall neglect of the Defence Forces left the country virtually at the mercy of both warring sides. It is likely that, had Collins lived, he would have been involved in the development of both military and civil aviation, a subject in which he had taken a great interest.

After the change of government in 1932, the tradition established in the 1920s of a post-manoeuvres review continued. In pouring rain the OC, Air Corps, greets Mr de Valera as he inspects the aircraft and troops in the Phoenix Park.

IX

Dublin 'Dromes – Baldonnel and Kildonan

THE SURVEY CONDUCTED BY MAJOR SHOLTO DOUGLAS in 1917 resulted in three Training Depot Stations being established around Dublin – at Baldonnel, Collinstown and Tallaght – as well as others elsewhere in the country. The Dublin aerodromes were all built to the same specification and each was designed to accommodate seventy-two aircraft. The basic plan called for six 'aeroplane sheds' or hangars, a repair shed, stores and accommodation for personnel. Baldonnel, about ten miles southwest of the centre of Dublin, covered just over 200 acres and came into operation on 1 September 1918. Two-and-a-half months later, when it had barely got into its stride, the Great War ended. Flying training was not crucial now and activities at all the training depot stations were scaled down. Although Baldonnel was still more active than the others, unrest among airmen waiting to be demobbed almost caused a mutiny there. The wartime ban on civil flying was lifted and the commanding officer at Baldonnel was given the following order:

> Detail forthwith one officer for civil aviation duty. This officer must be selected with great care. He must be a pilot or an ex-pilot and, in addition to possessing the necessary qualifications for supervisory work of this nature, must be possessed of great tact as the success of civil aviation depends largely upon initial treatment by the RAF. The officer selected must be ready to take up his duties on Thursday, 1 May, 1919.

A Captain MV McKeown got the job and his first customer was none other than Gordon Selfridge, whose company had recently acquired Brown Thomas's department store. Captain McKeown handled the paperwork, which included paying a landing fee of five shillings, £1 for hangarage of the DH9B for two nights and five shillings for a motor tender to Dublin. Incidentally, the charge for a motorcycle with sidecar for this trip would have been three shillings and four pence. The charter company which Selfridge used was Aircraft Transport & Travel (AT&T) which was the first commercial operator into Ireland. The company operated a fleet of sixteen DH9s and considered the Dublin route of sufficient importance to warrant the production of a special brochure.

About this time the first shots of the War of Independence were being fired and the military life of the aerodrome was reactivated with the arrival of Number 100 Squadron, RAF. As we have seen, the end of this conflict and the beginning of the Civil War saw the establishment of the Irish Air Service as part of the new National Army. This inhibited civil aviation, but resulted in occasional flights chartered by the British press to cover the conflict. The civil visits were very informal and the new Provisional Government viewed them favourably, particularly as one of the leading UK carriers of the time, Instone Air Line, was planning a London/Dublin link.

Instone considered the 'Fifteen Acres' in the Phoenix Park on the edge of the City to be the best location (it had been used as a temporary landing ground by the RAF as they finally departed from the Free State), and it hoped to start up as soon as the country was at peace. In the event, Instone never attempted a proving flight and quietly dropped the project.

In April 1923, at the official end of the Civil War, *The Irish Times* chartered another DH9 to fly photographs of the wedding of the Duke of York (later King George VI) to Dublin. The photographer also acted as courier and gave the following account of his trip: 'The wedding was at 11.30 and by 1.45pm I had taken all the pictures and got hold of the official photographs of the actual ceremony; then I was driven to De Havilland's in northwest London and boarded a DH9 which took off in a hail storm at 2.30pm.' After taking off from the De Havilland airfield, a headwind slowed the aircraft down on its way to Chester, which was reached at a few minutes past five o'clock; thence to the jumping-off point at Shotwick for a quick refuelling and on to a landing at Baldonnel at 7.35 pm – giving a trip of five hours and five minutes. In better weather

conditions the pilot reckoned he could have cut one-and-a-half hours off
that time.

Civil aviation was virtually dead until the end of 1925, when the
prototype of De Havilland's famous light aircraft, the DH Moth, arrived.
The pilot was the son-in-law of the noted artist Sir John Lavery, and his
companion was the British Director of Civil Aviation, Sir Sefton Brackner.
They had high hopes of selling the inexpensive trainer to the Air Corps
and any other takers. The sales trip was a success, and the Air Corps
ordered four DH 'Cirrus Moths'. Sir John Lavery captured their
departure for Baldonnel in oils. Commercial and private flying was now in
the ascendant, and accordingly Baldonnel officially became Customs
Terminal Aerodrome for Ireland. Here, visitors could clear Customs, be
refuelled and avail of hangarage and servicing facilities if required.

A chartered De Havilland DH50 operated by Imperial Airways
brought another salesman, this time from the British & Irish Petroleum
Company. His mission was to capture the lion's share of the country's
petrol requirements. This flight from Croydon actually missed
Baldonnel, as the east coast was shrouded in a heavy mist, and the DH50

A soft touch-down! Intent on scaring off straying sheep at Baldonnel,
the pilot of this amphibian forgot to lower the wheels, but no damage was done
to aircraft, pilot or sheep!

went on to the Curragh; an Air Corps machine was dispatched to escort it to its proper destination. Navigational technology in those days was limited to the 'Eyeball Mark I'. Another Imperial Airways DH50 turned up carrying urgent correspondence held up by the general strike of 1926 in the UK, and one more brought a couple of English jockeys for the Irish Oaks race meeting. On this occasion the Curragh *was* the predetermined destination, and the aircraft was an added attraction for the racegoers.

Shortly afterwards, Baldonnel received its first woman pilot. She flew an Avro Avian light plane, a strong contender, like the DH Moth, for the expanding private market. Mrs Elliot Lynn (later to become Lady Heath) was not just a weekend flyer, but was at the time the only woman in the world to hold a commercial pilot's licence. She had earlier covered 1,300 miles in one day, between three o'clock in the morning and 9.30 at night, making a stop at almost every English aerodrome and a total of eighty landings. Nevertheless, she too missed Baldonnel and landed in the Phoenix Park. She was escorted from there by Commandant James Fitzmaurice, flying one of the Air Corps's new DH Moths.

Another titled woman, Lady Abe Bailey, a well-known competitor in air races flying her DH Moth, made sure that her sea crossing would be uneventful by converting the passenger's cockpit into an extra fuel tank and equipping herself with a motorcycle tyre to act as a dinghy should she alight on the sea. She had flown casually over in one hour to visit the Horse Show and afterwards collected her aircraft at Baldonnel where Fitzmaurice, now a Colonel and famous for his Atlantic crossing, gave her a flight plan for her return, and gallantly escorted her out to sea.

In the summer of 1927 Colonel Frederick Minchin, who had earlier flown Imperial Airways aircraft to Baldonnel, arrived back in a private capacity accompanied by another experienced pilot, Captain Leslie Hamilton. They were flying a Vickers Viking, the first amphibian aircraft to visit the aerodrome. Their ultimate goal was the £5,000 prize money on offer from a Canadian brewery to the first crew to fly from London to its namesake – London, Ontario. They took a look at Clifden, County Galway, as a possible intermediate stop (the prize did not stipulate a non-stop flight), and then departed for Amsterdam, where the Fokker aircraft they proposed to use for the attempt was being prepared. Though their attempt was a serious one, they were a merry crew, who insisted on wearing sailors' hats and chucking out an anchor as their landing run finished!

Fitz (in 'mufti' on the left) greets Colonel Minchin and Captain Hamilton who arrived at Baldonnel in the summer of 1927 in their Vickers Type 67 Viking amphibian G-EBED. The pair were surveying likely starting points for their Atlantic attempt in a Fokker F VIIa owned by Princess Alice of Lowenstein-Wertheim. The princess, Minchin and Hamilton made their attempt in August of that year and were lost over the Atlantic.

They had decided to drop Clifden as a stopping point in favour of Baldonnel, but in the event, when they eventually set out for Canada, they overflew Ireland. Their aircraft was a Fokker FVIIA, registered G-EBTQ and named *St Raphael*. The colonel and the captain were accompanied by Princess Alice of Lowenstein-Wertheim. The *St Raphael*, which took off from Uphaven in Hampshire, crossed over Wexford and flew on over the Aran Islands. A garda in a Connemara village reported the last sighting of the aircraft, for it was never seen again – the Atlantic had claimed another victim.

Baldonnel was in a very rundown state at this time – one pilot who slipped on the muddy surroundings was hospitalised for a fortnight with an injured leg. Shortly after this mishap, improvements were reported:

> Whoever is responsible for the 'brighter Baldonnel' movement deserves the congratulations of all who have occasion to use the aerodrome. The road through the camp has been made up and rolled, and all that is asked of visitors is that they should drive at a reasonable speed, as there have been several complaints lately. Flowerbeds have been arranged and the whole aerodrome is assuming an air of dignity instead of a ramshackle collection of buildings and meadows. The fencing has been renewed, and a well-built gateway now leads to the aerodrome from the Air Corps's hangars. At long last Baldonnel is beginning to look like an aerodrome!

A new mobile floodlight was acquired but it would only be available for night flights by civil aircraft in the case of an emergency. The Air Corps

Big Bird at Baldonnel: On 1 July 1932, Helena, undoubtedly the biggest biplane ever to visit, was on a financial mission. When the British Government decided to redeem the entire issue of 5% War Loan stock, replacing it with a new stock yielding only 3½%, the forms for 50,000 disgruntled punters in Ireland had to be swiftly delivered and returned to London.

was still discouraging any idea of making Baldonnel a permanent civil airport. Nevertheless, towards the end of 1932 there was a steady influx of cross-channel air taxis bearing news reporters and newsreel cameramen to record the Irish scene. One taxi company lowered its rate to sixpence per mile on this route, half the price charged by virtually all its competitors. Feasibility flights continued to investigate the possibilities of scheduled air services linking Dublin, London and the Continent, but, though these proved positive, there was no governmental support and the trials petered out. Nevertheless, by the mid-1930s one could fly Baldonnel/Liverpool return for five guineas, courtesy of the Midland & Scottish airline, though this offer only lasted for six months.

The most significant development of this time was the founding of Aer Lingus and the beginning of scheduled services. Wireless and meterological services were laid on and a Customs hut and a waiting room for passengers were erected.

Hugh Cahill, who operated Iona Motor Engineering Works at Glasnevin in Dublin, was an enthusiastic member of the Irish Aero Club. He bought some land at nearby Kildonan and proposed to use it as a base for air taxis. These would initially serve Irish locations and London, and later Paris. He had an agency from the French manufacturers Desoutter, from whom he acquired a Mark I high-wing cabin monoplane; this was preceded by a Gypsy Moth which Cahill operated from Baldonnel prior to the opening of Kildonan. The Desoutter was damaged during a delivery flight but was soon replaced by a Mark II, registered EI-AAD and bearing the legend 'Iona National Air Taxis & Flying School'. In 1931 Iona moved to Kildonan and two more Gypsy Moths were acquired. *The Irish Times* reported:

> Mr Cahill has selected a fine field which, with some alterations which he proposes, will make an excellent aerodrome in which big machines will be able to land and take off. He has constructed a hangar that can hold twenty aeroplanes, in addition to a large clubhouse where ground instruction will be given to members of the flying school.

The Iona Moths toured the country on charter work and 'barnstormed' by giving local joyrides and instruction. A 1932 poster announced:

> Joyriding daily from five shillings per flight (longer flights on application); Air Taxis always available at one shilling per mile; Aerial

photographs of towns, factories, docks, etc; Dublin's Own Air Park is only four miles from Centre of City; Bus from Eden Quay; Tea Rooms for visitors, moderate charges.

Kildonan Aerodrome in its heyday, with Moths on the ground and a Monospar taking off.

Kildonan was now licensed by the Department of Industry and Commerce as a public aerodrome with customs facilities; aircraft from overseas could now choose between Baldonnel and Kildonan. Iona purchased an enclosed DH Fox Moth, EI-APP, which could take three to four passengers, and this became Ireland's principal charter aircraft for the following six years. Aircraft numbers increased; the original hangar space was more than trebled and workshops added. But nevertheless in 1933 Iona decided to quit the aviation scene, at least for the time being. Emerson Flying Services now took over the aerodrome, including Iona's aircraft and staff, and continued operations until the beginning of 1935. This company prospered and had a very good relationship with Aer Lingus, who were happy to loan staff in the cause of promoting aviation generally. Emerson Flying Services was headed up by Lady Cathleen

Emerson and it was succeeded by Dublin Air Ferries, the brainchild of her adventurous friend Lady Mary Heath.

Lady Mary was a remarkable woman who, as a member of the Irish Aero Club, had often used Kildonan. She started life in County Limerick as Sophia Catherine Pierce. She graduated from the College of Science, later part of UCD, and aged twenty-six won the women's world high jump record. In the late 1920s she established the world altitude record for a light aircraft – 23,000 feet – and she became the first woman to make a parachute jump. In 1928 she flew her Avro Avian from Capetown to London, but in the following year an air crash in America virtually finished her flying career. She was much married but kept the title Lady Mary Heath. Despite business commitments she found time to set up the Irish Junior National Aviation Club to encourage young people.

Her new company had three Gypsy Moths and a Fox Moth. The impact of regular Aer Lingus services affected her operations but she kept

In 1938, an Irish-American, Douglas Corrigan, was unable to get permission from the US authorities to fly the Atlantic in his tiny Curtiss Robin monoplane. He said that he would return to his Californian base, but instead he turned up, twenty-eight hours later, over Ireland and was escorted to Baldonnel by an Air Corps machine. On landing, he asked innocently, 'Is this Los Angeles?' His maps of the North Atlantic, which he had discarded just before landing, were picked up by an Air Corps corporal. The youthful aviator was instantly dubbed 'Wrong Way' Corrigan.

going until her remaining aircraft, the Fox Moth, had to be disposed of on 31 August 1938. Kildonan itself was closed at the end of that year and *The Irish Times* had the last word:

> During the past seven years, the various organisations operating from Kildonan trained a number of pilots and carried out many charter flights without any accident causing loss of life.

Iona restarted after World War II at Dublin Airport with the founder's son Pearse deeply involved in the business: he secured the US Cessna agency for Ireland and even briefly revived the Irish Aero Club. By the end of the 1980s Iona was maintaining forty aircraft in addition to its own fleet of seventeen. The legacy of Iona's policy of buying new Cessnas, operating them for a couple of years and then selling them on to other aero clubs around the country at a reasonable price can be seen in the predominance of the type in Irish airspace.

Tragically, Lady Mary Heath died in her early forties in a London bus accident. As she had wished, her ashes were scattered over her native Newcastle West. In 1999 the Irish postal service, An Post, issued a fine philatelic set commemorating famous Irish aviators. These included Lady Mary, a very fine Irish aviatrix whose many accomplishments included being the first woman to hold a commercial license in these islands, and probably worldwide.

X

TRANSATLANTIC ADVENTURES

FLYING THE NORTH ATLANTIC TODAY is a comfortable experience, made all the more pleasant in first class on a supersonic aircraft. Seven hours in tourist class is not too daunting either. Up ahead on the flight deck, the pilots can rely on an abundance of power and navigational aids, which can tell them within a matter of metres exactly where their aircraft is at any second. However, 33,000 feet below, the ocean may be raging, with howling headwinds of 100 miles an hour and gigantic waves. This was a frightening scenario for early aviators. Yet low-level flight was often forced upon them by the winds, particularly on flights from Europe to North America. The transatlantic traveller today has little idea of the trials of these trailblazers.

Between the world wars, long-distance flying was a daunting and often foolhardy endeavour. Why did so many attempt the North Atlantic crossing? In the words of a famous mountaineer when asked about climbing Everest, 'because it is there'! The distance from Ireland to a flat piece of dry land in Newfoundland is over 2,000 miles, which, at a speed of 100mph, will make twenty hours in still air. But rarely is the air still! Strong winds could force an early aircraft to stand still in the air or indeed blow it backwards; as for navigation aids, there were virtually none.

Crossing the Atlantic by air to join the two continents was a dream, partially realised at first in 1919 by a group of four United States Navy flying boats which 'island-hopped' from America via the Azores and Portugal, with one of the group reaching Plymouth in England. Later that year a non-stop crossing by Alcock and Brown from Newfoundland to Galway advanced the cause. Their aircraft was a converted wartime Vickers Vimy bomber, which, aided by following winds, made the crossing in sixteen hours. It was an outstanding achievement in the sixteen-year history of powered flight, and it foreshadowed Ireland's potential importance on this great air route. The spot where the Vimy landed is marked by a memorial shaft erected by Aer Lingus in 1959.

However, the mighty ocean could still not be flown from Europe to North America, and for a combination of reasons. The prevailing northwest winds, which gave a favourable tail-boost to a west–east passage, became treacherous headwinds on a reciprocal crossing. In addition, if, to avoid the persistent Newfoundland fogs, a more southerly route was chosen, the receding North American coastline would add on more miles of sea before a landfall. The immutable geographical 'catch twenty-two' would call for a greater fuel load and would add to crew fatigue; each extra mile could spell the difference between success and a watery grave. Conversely, a northerly course, devoid of accurate navigational fixes (compasses become unreliable in the polar latitudes), could delude a pilot into flying on and on into the uninhabited wastes of Labrador, and certain disaster. The dice, therefore, were heavily loaded against a successful crossing from Europe.

In the nine years that had elapsed since the first non-stop west–east flight, the Atlantic had taken a heavy toll of lives. In a dozen serious attempts on the crossing from Europe – ten of which were made in 1927 – seven fliers had been lost, and both officialdom and public opinion opposed 'further costly life-threatening adventures'. But the Atlantic airspace had to be tamed. In one attempt, a pair of German aircraft, which had started from the Fatherland, covered a distance of almost 1,000 miles before storms forced them to return, despite being in the air for twenty-four hours.

There was at least one Irishman who dreamed of success – James Fitzmaurice, better known as 'Fitz'. He had joined the infant Irish Air Service in 1922, and commanded the Fermoy base during the Civil War. By 1927 Fitz was the Air Corps OC. An experienced pilot and navigator, he was the epitome of the soldier of fortune. His ambition was to make a solo crossing in *The Big Fella*, by converting its passenger compartment into a large fuel tank and fitting a bubble-type canopy over the open cockpit. In the course of many tests and discussions, Rolls Royce reassured him of the reliability of its Falcon engine. Fitz got together a sum of £1,000, which he offered the Department of Defence for the machine, but this was turned down.

Lahinch, 1936: Pond and Sabelli, en route from New York to Rome, made this unscheduled stop in County Clare. Sabotage was suspected, but Air Corps mechanics got them going again.

In the same year a Scot, Captain RH MacIntosh, an Imperial Airways pilot (better known as 'All Weather Mac' because of his tenacious spirit), flew into Baldonnel in a Fokker F VII. This was registered G-EBTS, and named *Princess Xenia* after the daughter of the King of Greece, who was married to the American backer of Mac's Atlantic attempt. MacIntosh was accompanied by a co-pilot, but trial flights made it clear that the effects of a war wound on this veteran's knee would prevent him from holding the rudder pedals for any length of time. Fitz readily stepped in to replace him! Continuous bad weather forecasts frustrated a take-off. The two fliers were extremely anxious to be on their way, because an American millionaire, Charles Levine, was poised to make the flight from England. Keeping an eye on the weather, Levine and other likely starters, MacIntosh declared that if anyone else started out they too would be off! Further urgency was added because MacIntosh's leave of absence was running out and Imperial Airways refused to extend it. He had in mind to achieve three records: the first east–west crossing; the longest flight, by returning over the outward route; and the first flight across the Atlantic and back in the same aircraft.

At last, on 16 September, the blue-and-silver *Princess Xenia* was airborne, with two Air Corps aircraft accompanying it. In poor visibility over the midlands, the army planes turned for home. As the Fokker pressed on, weather conditions continued to worsen. By the time the pilots were 300 miles out over a raging ocean, they suffered severe

turbulence and were worried by the rough note that the engine had developed. To return became the only option, and that in itself proved hazardous, as Fitz plotted a course that would bring them south of Galway. To lighten the load he began to jettison fuel but, as he was checking the rear fuel tank, he was blinded by the last gush of petrol. Rinsing his eyes, his sight gradually returned, and in great pain he managed to struggle forward to assist MacIntosh in steadying the aircraft.

'Fitz' and 'All-weather Mac', dauntless pilots of Princess Xenia, were to meet many years after their transatlantic adventure as guests on 'This is Your Life'.

The battered *Princess Xenia*, having made the planned landfall, was turned southwards until Fitz spotted a shingle beach in Kerry where it put down safely. The tide was coming in but there were many helpers to pull the aircraft above the high-tide mark and secure it. Thus ended yet another unsuccessful attempt to challenge the mighty force of the Atlantic – but seven months later Fitzmaurice was to pick up the gauntlet once again.

The German crew, who had spent twenty-four hours attempting the crossing from their base at Dessau, now planned to make Baldonnel the starting point for a second attempt. This would cut the distance on a flight where every mile and every gallon counted. Once again they flew in *Bremen*, the single-engined all-metal Junkers W33, registered D-1167, owned by Baron Von Hunefeld, a non-pilot. His companion was Captain Hermann Kohl, a pilot of great experience. Aware of Fitz's courage and experience and of the fact that he commanded their new starting base, they were delighted when he accepted the challenge to accompany them.

Bremen: A Junkers W33 Freighter. Wingspan 70.75m, length 10.5m, speed 180km/h, range 1000km. The type served with Lufthansa from 1928 to 1942.

The flight of *Bremen* has often been described, and here we will dwell only on some of its highlights. As *Bremen* took off from Baldonnel's rain-soaked grass a sheep strayed into its path and Fitz hauled back on the controls to avoid hitting the animal. *Bremen* bumped down again but slowly rose in a flat curving climb, narrowly missing some trees. The Air Corps's lone Fairey IIIF patrol plane also took off and escorted the adventurers until they disappeared into the mist around Athlone. The two pilots flew on steadfastly, nourished by soup and sandwiches provided by Von Hunefeld. For ten hours, 'Sunshine and peace reigned everywhere.' But over the ocean the going got tough. To calculate the strength and direction of the contrary winds, smoke bombs were dropped to make drift readings to keep *Bremen* on track. The aircraft was brought down to sea level to minimise the headwind effect, and when the wind abated it rose to 300 feet.

At the supposed halfway mark continuous icy cloud persisted; and when the crew reckoned they were only three hours from land they actually had a tough twelve hours in darkness, followed by fog, still ahead of them. The night-flying skills possessed by both pilots were tested to the full as the steady diesel engine droned on. A storm erupted, during which Fitz noticed oil on the cockpit floor. 'I opened the roof of the cockpit cabin and, standing on my seat in the slipstream, I pushed myself between the two large petrol tanks in the rear and eased myself forward until I could get my head under my control column where the oil tank was located.' Luckily the problem proved to be a minor one, but it took Fitz an hour to solve it.

This scare caused the airmen to change course in order to reach land as soon as possible. But the violent storm had thrown *Bremen* off course, and continuing turbulence and changeable strong winds had the small aircraft at their mercy. At long last, *Bremen* broke out of the fog and into calmer, starlit skies. A couple of hours later, Fitz spotted Labrador's Torngat Mountains. Soon, however, the coastline was obliterated by snow showers, but as Fitz recollected:

> Suddenly, the curtain of snow lifted and we observed in the distance the outline of what we believed to be a large ship, frozen in the ice … what a tremendous relief! It seemed an eternity since we waved a long farewell to the lighthouse-keeper at Slyne Head in Galway.

The three pioneers being received by the exiled Kaiser. The Irish authorities were dubious about the political implications of a serving officer having an audience with the deposed emperor. Von Hunefeld took this opposition as a personal slight – Wilhelm II was still his emperor – and the diplomatic ramifications were forgotten.
(Fitz is on the right of the Kaiser.)

When Europa, sister craft of Bremen, was due to arrive with the victorious trio at Baldonnel, fifteen Air Corps planes, virtually the entire airforce, escorted it. Here the escort is about to take off to greet the returning heroes.

After the landing at Tempelhof, Berlin's airport, Fitz and the baron poke fun at Kohl, who is suffering from ear trouble. Meanwhile, the band plays on, on the right.

The 'ship' turned out to be another lighthouse, standing on Greenley Island and close to a frozen freshwater reservoir. On this Kohl made a marvellous landing, the only damage being a bent propeller and undercarriage strut. The unforgiving North Atlantic had at last been fully conquered, and the whole world was quick to recognise the feat.

Naturally, Ireland and Germany were exultant, particularly the latter, the country now rejoicing in its first post-war success.

Honours were heaped on the crew during a lengthy 'lap-of-honour'. Fitz was promoted to the rank of Colonel, but resigned from the Air Corps shortly thereafter and spent much time abroad. His optimistic plan for a solo Atlantic crossing never came to fruition, and he drifted. The international renown that he had brought on his native country, only six years after it had become independent, was never fully recognised at home. In 1965, when the remains of Sir Roger Casement were repatriated via Baldonnel, the Fianna Fáil government decreed that the base would henceforth be known as Casement Aerodrome. Perhaps under another government it would have been named 'Fitzmaurice Field', and Fitz would have deserved it, because he was Ireland's greatest aviator.

Fiftieth Anniversary Celebrations, 1978: the General Officer Commanding, Air Corps, Jerry O'Connor occupies the seat in Bremen where his predecessor Colonel James Fitzmaurice had sat on the record flight. This photograph was taken in the Ford Museum, Michigan, where the aircraft was displayed until its recent return to Bremen Airport.

XI

THE IRISH AERO CLUB 1928-1939

THE FIRST IRISH AERO CLUB was born in November 1909. It soon changed its name to the Aero Club of Ireland, and then faded away when the Great War broke out in 1914. When peace was restored, private flying soon became popular again in the UK, aided by government subsidies and by the appearance of light club aircraft – the Avro Avian, the Blackburn Bluebird and the legendary De Havilland Moth. In 1925, a number of air-minded people got together in the Royal Hibernian Hotel in Dublin. It was brought to the meeting's attention that, though there were at least sixty flying clubs in Britain and two in Northern Ireland, there was none at all in the Free State. However, only thirty people at the meeting showed any real interest; at least four hundred paying members would have been required for the purchase of a light aircraft.

Spurred on, no doubt, by the success of *Bremen*'s transatlantic victory, the enthusiasts regrouped in 1928. A committee was appointed with Colonel Fitzmaurice, the conquering hero, in the chair. A publicity campaign was launched, inviting people to join a proposed Civil Aviation Development Association, a form of holding company for a proposed club. Avro despatched one of its Avians and an advanced version of its sturdy 504K trainer. Demonstrations were given at Baldonnel and the Phoenix Park and over fifty 'serious' prospective members were taken up for what, for most of them, was their first flight. So it was that the Irish Aero Club was reborn on 15 August 1928, with a paying membership of over 150, on the strength of which an Avian was ordered at a cost of £750.

One of the new members ordered his own Avian and both aircraft arrived at Baldonnel in the autumn. The purchases had the distinction of being the first civil aircraft to be registered in the state, and the club

DH 'Moths' in company. The first two aircraft on the Irish Civil Register, EI-AAA and EI-AAB.

aircraft bore the registration EI-AAA, the private machine being allotted EI-AAB. To copperfasten the whole enterprise, the Club became a limited liability company.

The Air Corps, not overburdened with military craft, was happy to accommodate the two Avians in a hangar, and donated a large hut, which was converted into a comfortable clubhouse. Although the Government was not prepared to offer any direct financial subvention, it permitted Air Corps technicians to undertake overhauls and repairs at a nominal charge. Air Corps pilots acted as instructors in their spare time, and when regulations prevented them from actually flying, their expertise at ground level was still readily forthcoming. Club members could train for a pilot's licence and then hire out its aircraft, while less daring types could become 'passenger members', to be borne aloft at special rates. Indeed, non-flying members were also welcome and no doubt contributed to the bar receipts! In addition to the Club's Avian, the owner of EI-AAB and another member who owned a Moth, EI-AAC, put their aircraft at the disposal of members when they themselves were not flying.

By 1930, the Club was firmly established and its activities were of great benefit to the beginning of Irish civil aviation. The government had yet to set up a licensing system, so pilots had to go to the UK to qualify, but this was resolved when the Air Navigation (General) Regulations were formulated. There were two licenses available: a Class A Licence for private flying and a Class B for those who wanted to 'ply for hire or reward'. After two years in existence the club had trained eight pilots to A standard, and there were now over 150 active flying members, who had amassed thousands of flying hours and much experience between them.

Oliver St John Gogarty, a man of many parts and a prime mover of the Club, was totally hooked on flying! He had forecast that Ireland would become a natural stopping-off place when commercial

'Picking up the Handkerchief' – with a nine-inch spike on the wingtip.
A great crowd pleaser at Irish air displays, which required great skill.

transatlantic services commenced. His description of his own first flight reads:

> The engine is already running ... we are blown by the slipstream as we climb into our cockpits which are the portal of the blue sky ... the roar of the engine increases, and increases your assurance ... the men leaning against the wings, and the man who leans on the tail, leave go. The wedges are pulled from under the wheels. Our speed increases. It is blowing over sixty miles an hour. Our tail is up. The seat which slanted back is level now. Vibration is intermittent. Now there is none at all. We are flying. No silence on earth is equal to it. It falls on the spirit softer than sleep. I can understand the trance that fell on Aucassin, though he was in mid-battle.

Dr Gogarty was convinced that, on purely medical grounds, the acceleration of flying was beneficial for many kinds of nervous cases, and he declared that middle-aged and elderly people would benefit from what he described as 'the marvellously rejuvenating effect'.

The Senator upended: Oliver St John Gogarty, on landing this Cirrus Moth at Baldonnel, encountered a sheep – a perennial nuisance. The eminent senator was not amused when this photograph appeared in the Evening Herald.

Gogarty – senator, writer and skilful surgeon – was much in demand in London and was allowed to hire Air Corps aircraft when he had to travel to perform operations there. The extremely high fees which he commanded paid for this transportation. Once, while a passenger in the Corps's single Fairey IIIF, he accidentally pulled the ripcord of his parachute and was almost dragged out of the plane and down into the sea.

Manco Scally of Tullamore was another enthusiastic member. He acquired a tiny Comper Swift in which he proposed to fly to Ceylon (now Sri Lanka), though he lacked the experience required for such a trip. Undaunted, he took off from Baldonnel in his green, white and gold monoplane, which he had dubbed *Shamrocket* – he had wanted to take off from central Dublin but the Corporation demurred. He got as far as Marseilles, but unfortunately, on coming in to land there, he hit some tall trees and was killed. Enthusiasm for long-distance flight has to be tempered with experience and harnessed to the right kind of aircraft for the job in hand.

Things were going so well that the Club bought a second aircraft, the most up-to-date development of the Moth, which had a metal fuselage, giving it the name Metal Moth, though officially it was a DH60M. By this time, nine private machines were owned by Club members, including a Puss Moth, a monoplane with an enclosed cabin in which the owner could smoke to his heart's content! Also in this number was another Moth, part of the Hon. AE Guinness's fleet, which also included a Saro Cutty Sark amphibian and a large three-engined Supermarine Solent flying boat, which he kept on the water at Cong in County Galway.

President of the Executive Council, WT Cosgrave TD in a Club 'Moth' with pilot Charlie Russell.

Thus far, apart from *Shamrocket* and a couple of small incidents with no damage to the fliers, the Club was free of accidents. Compared to the cost of private flying today, it was much more easily affordable to take to the air in the 1930s. Full annual flying membership cost £5, and the cost of dual tuition per hour was thirty shillings (£1.50), and five shillings (25p) for a solo trip; therefore, with an average of fifteen flying hours required, a member could gain a licence for little more than £20. Gliding also was proposed for those who fancied the attractions of engineless flight. Charles Russell, who had left the Army, and his brother Arthur were honorary instructors, but now it was decided to appoint a full-time Chief Flying Instructor (CFI).

'Open Day' 1931 at Baldonnel: a highly successful event, organised by the club.

In the summer of 1931 the Club organised its first annual air display. It was held at Baldonnel over a three-day period in August and was a tremendous success – which probably helped to provoke the parsimonious government into making a grant of £1,000.

By the beginning of the next year, the Air Corps handed over the 'club' hangar entirely for civilian use – it already housed five private machines. With the arrival of a full-time CFI, flying training was constantly on tap, weather permitting. Club members gave lectures each week at Bolton Street Institute of Technology, covering such subjects as engines, map reading, rigging and meteorology. Dances and other social events were also held frequently throughout the year. The emergence of Ireland as a separate nation caused difficulty when the Club initially applied to the International Sports Flying Federation, which thought that the Royal Aero Club of Great Britain was the appropriate governing body for the Free State!

After nearly five years of safe flying, the Club had its first fatal accident (in fact the first to a civil aircraft in the Irish Free State) when its Gypsy Moth EI-AAH inexplicably stalled and plunged into the sea near Dalkey in May 1933. The pilot was giving a friend an aerial view of Dublin Bay and its environs when the accident occurred. The publicity given to it did nothing to help the cause of private flying. The Club ordered a replacement Moth, demonstrating its complete confidence in this excellent little aeroplane. At about this time, a branch of the Club was formed in Cork.

The Club, in collaboration with the Air Corps, enthusiastically took on the arrangements for the King's Cup air race of 1937, a two-day affair with Baldonnel as the overnight stop. De Havilland's at Hatfield, Hertfordshire, was the starting point where twenty-seven eager competitors took off on a fine Friday morning. At Scarborough, the first turning point, a fatal accident occurred, but the race continued, though several of the competitors were forced to drop out.

Twenty-one aircraft landed at Baldonnel, where a public address system had kept the thousands of onlookers fully informed throughout the race. One of the competitors failed to cross the finish line properly and had to take off again and make a correct landing – by which time a rival had slipped in ahead of him! The public enclosure gave an excellent view of all of the different aircraft being refuelled and adjusted in readiness for the following morning. That night, the Club entertained the competitors, with dinner at the Shelbourne Hotel in Dublin. One of the guests noted that the Irish hosts and

The King's Club Air Race 1937, when Club and Air Corps organised the overnight stop at Baldonnel.

the general public referred to the competition as 'The Big Race', rather than mention His Majesty's association with it. The new Irish Constitution, which would get rid of him, was about to be launched!

Next morning, the field was down to seventeen – four had been eliminated and another four had had to drop out during the second leg, leaving Charles Gardner in his Mew Gull racer to finish first at Hatfield.

He had done an average speed of 224 mph over the 1,442 mile circuit.

The Club's arrangements for the race, a high point in each aviation year, were highly praised, but the expenses involved added to its financial difficulties. For a number of years the balance sheet had not been in the Club's favour – it now owned only one aircraft. As winter approached, further limiting fee-paid flying, a receiver was called in. The State claimed that it was a preferential creditor (it was owed about £300), but the liquidator rejected this and the matter was referred to the High Court in a case which became a legal precedent. The outcome was that the State was treated like other creditors and, like them, received a dividend in the pound. Students of constitutional law will remember the Irish Aero Club long after others have forgotten it!

The Club had lasted for almost a decade and, even without money problems, it would have had to close down, like its predecessor in 1914, on the outbreak of World War II, when all private flying was stopped. Afterwards, flying clubs began to emerge at local level, albeit slowly, and the title Irish Aero Club, though briefly revived, soon became just a reminder of the halcyon days of pre-war flying.

The Cobham Air Circus comes to town! The Club also participated in this round-Ireland tour.

XII

VELVET STRAND: IRON MEN

ONE GLORIOUS JUNE DAY IN 1930, a strong-faced Australian pioneer pilot was playing Portmarnock golf links when the long stretch of nearby beach revealed itself. He turned to his companion and asked: 'How long has this been here?' His opponent replied: 'It isn't always like this; the tides and currents can change what you see into a most unsuitable runway.' The Australian was Charles Kingsford-Smith (universally known as 'Smithy'), a veteran of the air war in France. After the war he had blazed a trail across the Pacific from California to his native land and the following year broke the Australia-to-England record in his tri-motored Fokker, appropriately named *Southern Cross*. He was now ready to follow the trail of *Bremen* across the North Atlantic and complete a circumnavigation of the globe.

'Swinging the Compass': Southern Cross at Baldonnel before its short hop to its take-off point at Portmarnock.

Baldonnel's longest run of one mile was still half-a-mile too short for Smithy's plane when heavily laden with fuel and a four-man crew.

Obar in aisge (a waste of time): These troops on the Curragh are busy preparing a runway for Southern Cross, before its pilot, Kingsford-Smith, decided instead on Portmarnock Strand, to which the troops cheerfully turned their attention!

With unstinting cooperation from the Army, he had been offered the use of the flat plains of the Curragh Camp as a runway, and troops were busily preparing the ground. However, Smithy now opted for the vision he had seen at Portmarnock beach, known as the 'Velvet Strand', and the hard-working troops made sure that rocks and other hazards were cleared away. Despite the good weather in Ireland, the forecasts for the Atlantic were uncertain and it was not until 22 June that a favourable Met report was received, though it did warn of dense fog banks from Newfoundland down to the State of Maine.

Smithy had earlier advertised for an experienced navigator and he found him in the person of Captain JP 'Paddy' Saul. Paddy was a Dubliner who went to sea when he was fifteen and had obtained his Master's Certificate in navigation by the time he was twenty. He joined the Royal Flying Corps in the Great War and, home again, he was on the committee of the Irish Aero Club.

In the middle are two of the crew of Southern Cross – Stannage the wireless operator and Saul the navigator.

Southern Cross at Baldonnel with the Aero Club Moth, EI-AAA, and an Air Corps Vespa.

The second pilot, Evert van Dyck, came from Dutch Airlines; he was a six-footer who contrasted sharply with the small South African wireless operator, John Stannage, a merry soul but a true professional. *Southern Cross* was flown over from Baldonnel and positioned on the strand where, despite a small break in the weather, thousands had gathered by 4am. The Air Corps support team had fuelled the aircraft with 1,300 gallons of aviation spirit, and, once the crew were aboard, had sealed its door and windows with doped tape to lessen the cold and improve the streamlining. When its three engines had been revved up to their maximum, *Southern Cross* slowly rolled down the compacted path of the Velvet Strand and was off! Ironically, despite 'flaming June', it encountered low cloud and rain and little was seen of the countryside until, at 800 feet over Athlone, visibility improved and soon Slyne Head appeared. This was Paddy Saul's navigational benchmark for the real start of the adventure.

Over the ocean occasional patches of blue appeared and, with encouraging wireless reports from ships further out, the crew, whose conversation was drowned by the roar of the engines, tucked into a good breakfast. With Paddy Saul monitoring their progress, they eventually spotted trawlers fishing on the Newfoundland Banks and, in celebration, a bottle of Jameson was broached, some of which Smithy, in his semi-open cockpit, used to ease his petrified cheeks! A different form of first aid was required by the South African who had received some nasty gashes on his hand as he oiled the bearings of his fan-operated generator.

Now Saul takes up the story:

> Midnight of the first day found us flying well, with the weather a little clearer, and again we went down closer to the sea to check our altimeters. Shortly afterwards we saw the stars for the first time, and I managed to get a couple of good fixes on Polaris and Pollox. These coincided nicely with the wireless bearings from two ships. Now, we were only 350 miles from Cape Race.

But one hour into day two, *Southern Cross* ran into a solid white fog bank and immediately climbed for safety. Even after it had passed 7,000 feet the fog was still impenetrable and, having regard for fuel consumption, the pilots levelled out. Suddenly, all four men were attacked by violent nausea and recurring fits of vomiting, probably from petrol fumes. As if that was not enough, the unceasing vibration had caused the pilots' compass to break loose. The master compass now started to go haywire, but luckily an earth-inductor compass seemed to be behaving properly, and Smithy reported that they were on a very steady course – until he discovered its needle was stuck! Saul divested himself of his flying suit and after four-and-a-half hours he repaired the earth induction compass.

Years later he recalled:

> While on this job I had time to think on the foolishness of men in general. Here was I, who might have been lying comfortably in bed at home, out in the wild wastes of the Atlantic feeling very cold, and thoroughly miserable as the bouts of vomiting continued.

Concurrently, John Stannage was hard at work with his errant wireless set, as it lit up the cabin with brilliant blue flashes. Soon the radio began to pick up signals from Harbour Grace aerodrome. The pilots were encouraged to creep lower until suddenly, under the nose of the aircraft, a line of black cliffs, surmounted by a lighthouse, appeared. The lighthouse landfall recalled that of *Bremen* two years earlier, though a proper runway awaited *Southern Cross*, vastly better than the frozen reservoir on which the German aircraft had put down. Soon, smoke signals and waving sheets indicated the wind strength and direction on the runway at Harbour Grace and, after a couple of passes, Smithy put *Southern Cross* down on firm ground. They had been in the air for thirty-one hours!

The King's Club Air Race 1937, when Club and Air Corps
organised the overnight stop at Baldonnel.

Two days of rest and relaxation followed, after which the airmen flew on to New York and then, by stages, across the great continent. Eventually, and not without some hair-raising experiences en route, they touched down in Oakland, California, where Smithy had started his round-the-world flight some two years previously. It seems appropriate that it was the fourth of July, even though the crew was composed of an Australian, a Dutchman, a South African and the Irishman who had guided them accurately all the way!

Paddy Saul's future career included passing on his navigational skills to the RAF before the outbreak of World War II. He introduced the practice of instructing members of the Women's Auxiliary Air Force. In mid-war he came home to set up and operate the Air Traffic Control (ATC) service at Foynes and when the flying boat era ended, he remained as ATC chief in the Department of Industry and Commerce until his retirement. After the flight of *Southern Cross* it became the fashion in some British circles to describe its transatlantic leg as the first *successful* east-to-west crossing, simply because *Bremen*'s undercarriage had been slightly damaged on landing. Even Kingsford-Smith succumbed to this myth, but not Paddy Saul, who always gave first place to *Bremen*. Greater damage was suffered by *Bremen*, but only when the US Air Corps later attempted to fly it down to New York. By the 'Imperial' criterion, the flight of Alcock and Brown in 1919 was unsuccessful too, as their Vickers Vimy broke its back when landing in Connemara!

The next attempt from Portmarnock was that of Smithy's great friend, Charles Ulm, who had accompanied him when they crossed the Pacific. Ulm arrived at Baldonnel in his Avro Ten, called *Faith in Australia*, but sadly he never got off the strand. When the fuel tanks of his aircraft were virtually full, it started to sink into the sand, and during a strenuous effort by the gardaí to move the machine onto some wooden planks, a tie-rod on the undercarriage suddenly snapped and the wing on that side collapsed onto the sand, injuring some of the helpers. The Air Corps had not been involved in the take-off preparations, but an SOS was sent to Baldonnel when the incoming tide overwhelmed the aircraft and with it all of Charles Ulm's hopes – everything he owned was tied up in *Faith in Australia*. The Baldonnel team siphoned off the fuel but the tide, rising once again, was soon lapping around the aircraft. A young NCO slashed both sides of the fabric-covered fuselage, thus allowing the water to flow through without twisting the structure, but the damage had already been done. Eventually, the team pulled the wreck above the high-water mark

Collapse of Faith in Australia; it lived to fly another day.

where temporary repairs were effected before it was returned by ship to its manufacturers. Through the good offices of a sponsor, *Faith in Australia* was ready to fly again by the end of August but, though Ulm was keen to return to Portmarnock, worsening weather ended the transatlantic flying season of 1933, and the plucky Australian looked elsewhere for other skies to conquer.

The Velvet Strand had shown that it had sufficient bearing for even a heavily-loaded large aircraft, and so it was with no qualms that Jim Mollison selected it for a solo attempt in 1932. He had earlier married Amy Johnson, world famous for her solo flight from England to Australia. Her husband's light aircraft, like hers, was from the De Havilland Moth stable and he called it *The Heart's Content*. Like all his predecessors, Mollison hoped to reach New York direct but, as with them, headwinds and consequent high fuel consumption forced him to land at Newfoundland. Nevertheless, he accomplished the first solo flight from Europe to North America.

Since the 1930s, successive winter tides and winds have altered the beach at Portmarnock, which is still called the Velvet Strand. Erosion has caused a rock formation to be exposed which, though it limits the strand's suitability as a runway, paradoxically adds to its natural beauty. Portmarnock is now a suburb of Dublin from where the city's international airport can be seen. Powerful jets now overfly the strand where once iron men on fabric wings ventured forth and pushed forward the frontiers of flight.

Above: Jim Mollison checks his plane at Portmarnock.

Right: (Left to right) Lady Mary Heath, Amy Johnson and her sister Eleanor watch Amy's husband, Jim Mollison, take off. A reporter is intent on their reactions.

XIII

Aer Lingus Takes Off

IN THE AUTUMN OF 1921, between Truce and Treaty, Michael Collins, in the hope of a new Irish administration, speculated on aviation both in war and peace: 'We cannot afford enough military aviation to meet our requirements, if we have to fight again.' He saw in civil aviation a reserve airforce for the Army, and apparently laughed good-naturedly when somebody pointed out that under League of Nations regulations, civil aviation must not be used for the development of military air operations – a pious hope as subsequent events proved in Germany. Collins set up an Air Council, he himself attending in his capacity as Minister of Finance. A National Air Policy was drafted, which included provision for a civil school of aeronautics, a school of flying, and the establishment of a daily air service between Dublin and London. The slogan which had rung down centuries of Irish history – 'Break the connection with England' – did not preclude an air link!

A Director of Civil Aviation was appointed, and the funding of both civil and military flying was discussed in some detail. Later on, during the less turbulent sessions of the Treaty negotiations, Collins insisted that an annex to the agreement should contain a clause stating: 'A convention for the regulation of air navigation between Ireland and Britain will be drawn up.' The implementation of this clause would not happen for many, many years.

Elsewhere in Europe, the development of air services began shortly after the Great War ended, with Britain concentrating initially on establishing links with the Continent. Ireland, meanwhile, was embroiled in a dreadful civil war and afterwards faced a major reconstruction programme. Various estimates of the cost to the country have been made, but a figure of £50 million, a huge sum then, is probably accurate. The 'burnt earth' left behind by the Civil War, combined with Ireland's traditional agricultural economy, low income by comparison with its neighbours and of course its isolated location on the western edge of Europe, all contributed to a slow start in civil aviation. Conditions were aggravated throughout the 1930s by a spiteful economic war with Britain, which did no good to either side. Setting up a scheduled air service to anywhere was a very low priority indeed!

Nevertheless, many prominent people advocated a national airline. In the mid-1930s the Government authorised the setting up of 'any international airway between the State and any other country with a view to the limitation or regulation of competition as may be considered necessary in the public interest'. At last Aer Lingus was formed, with the help of Blackpool & West Coast Air Services which provided the money, technical assistance and an aircraft, a De Havilland DH84 'Dragon'. The service was marketed as Irish Sea Airways, and the first route was Dublin/Bristol which was opened on 27 May 1936.

Aer Lingus's first aircraft, which constituted its entire fleet in 1936, Iolar ('Eagle'), a De Havilland DH84 Dragon 2. The Dragon inaugurated the company's first and only route, from Dublin to Bristol, on 27 May that year, carrying a full load of seven passengers!

Next, a Liverpool/Isle of Man/Dublin service was inaugurated, mainly with Dragons, which were reinforced with four-engined DH86s during the summer months. The Irish partner also started its own independent summertime service to the Isle of Man. By September, the first air link, albeit an indirect one, between Dublin and London was established by extending the Bristol route to Croydon. A newly delivered DH86 Express Airliner came on stream and the Dragon was utilised on a new Dublin/Liverpool service. During these developments the British partner continued its cooperation and set up its own service between Croydon and Baldonnel, using DH86s and the newer DH89 Rapides.

The airline had to be an *Irish* one to qualify for Government help. In the Civil Service, there were divisions about ownership – one group visualised a publicly-owned airline while another wanted a state-owned company. The matter was settled in the following spring when a government holding company, Aer Rianta, was set up to buy the operating company and, retrospectively, become the owner of Aer Lingus.

The tempo of life of Baldonnel was quickening as civil operations started up, but the Air Corps languished as the Defence Forces remained at a very low ebb. Improvements at Baldonnel were made to the 'upper camp', the area around the hangars, and elsewhere. However, it remained very much a military station, and the company secretary of Aer Lingus found it necessary to write to the Officer Commanding as follows:

> It was mentioned to me that some of your military police on the gate have a rather narrow idea as to who should be let into the camp. They maintain that our staff – Station Superintendent, his assistant, Ground Engineer and his assistant, etc. – should have passes to go in and out, and that only passengers who arrive on the Dublin Tramways bus are entitled to go up to the office, unless they have tickets to show that they are booked on a plane. My view is that, where you have a sentry on the gate and a large notice outside it, no one who has not some real business would have the courage to ask to get in.

The airline's DH84 Dragon, was housed in hangar space leased from the Corps, and maintained by its chief technical NCO, who had transferred to the civil side. As the company expanded, it sought further personnel from the Corps: a senior officer to become manager, and pilots to join the civilian aircrew already employed. The Air Corps was by now re-equipping slowly and the Commanding Officer responded: 'As you are aware, the Corps is very short of pilots, and it is only with great difficulty that we are carrying out our own expansion programme. We can ill afford to lose the services of any pilot. However, bearing in mind the necessity for co-operation between military and civil aviation, we will continue to assist as far as we possibly can.' As a *quid pro quo*, the CO made it quite clear that he expected the airline would give preference to Air Corps reservists when filling vacancies – as it did.

All the Aer Lingus machines were fabric-covered biplanes, but in the last spring of peacetime the company bought a pair of state-of-the-art aircraft, all-metal Lockheed monoplanes from America – just in time for services to be curtailed! When war broke out, all cross-channel services from Dublin were suspended, although two months later a limited Liverpool service was restarted. But the war seemed far away and, as 1940 dawned, Aer Lingus transferred its base to Collinstown, an old RAF aerodrome. This was now revitalised as the new Dublin Airport, centred on architect Desmond FitzGerald's award-winning terminal building which was nearing completion.

The Lockheeds were sold because of restricted travel, but a pair of Douglas DC3s was on order. These had been shipped from the US to the Fokker company in Holland, who carried out DC3 assembly for the European market. The DC3, a type which first took to the skies in 1936, was to become the most popular piston-engined passenger type of all time – they are still flying today. The first Irish DC3, EI-ACA, arrived at Dublin Airport in January 1940, finished all over in Holland's national colour, orange, to identify it as neutral in the warring skies. In the event, the second plane didn't make it through to Dublin before the German Blitzkrieg overran the Low Countries. As events transpired, it is unlikely that Aer Lingus could have found work for the second DC3, so curtailed were its activities now. Incidentally, Swedish armoured cars that had been ordered by the Irish Army for shipment via Holland met the same fate as the second DC3, and were absorbed into the German forces.

Nevertheless, one state-of-the-art aircraft was better than none, and EI-ACA was put into service immediately. The orange enamel was slowly chipped away to reveal gleaming aluminium until eventually its piebald appearance caused it to be taken out of service and given a good going over with paint-stripper. It didn't stay long in its silver livery because 'total war' caused it to be camouflaged and plywood blackout covers to be fitted to its windows so that passengers could not observe military moves in the UK. This caused a very claustrophobic feeling, but did not prevent customer numbers from building up. A pair of Wright

Aer Lingus's first DC3, in World War II camouflage.

Cyclone engines powered the aircraft and a representative of the engine company would occasionally visit. Apparently, due to wartime clothing shortages, the representatives had one decent green suit between them which, whether the wearer was 5ft 2ins or 6ft 2ins, was always trotted out until it eventually became a threadbare relic. But at least the colour was right!

The airline offered its redundant DH86 to the Air Corps, who turned it down flat. The Corps, which was desperately short of coastal patrol aircraft, felt that they should have been given first refusal on the Lockheeds. A military version of the type, the Lockheed Hudson, had been acquired by the Corps via an RAF forced landing, but the hope of a three-plane flight of these excellent aircraft to reinforce its Ansons was dashed.

Aer Lingus settled into its new home at Collinstown's not yet completed terminal buildings and housed its five aircraft in a small hangar – a more commodious one was still incomplete due to wartime shortages of steel. From here the DC3 and other types set out on their daily schedule to Liverpool – often to be diverted to Manchester when the Luftwaffe was pounding Merseyside. The only problem with the early DC3 was a vicious steam-heating system which would either boil or freeze the passengers. In the latter situation, the passengers at least had specially woven Donegal tweed rugs – from the start the company was developing an Irish image! Its pilots were given strict instructions to avoid convoys in the Irish Sea lest they be fired upon by trigger-happy gunners. In fact this danger was closer to home: aircraft could only fly in Dublin airspace with prior express permission; otherwise they would automatically be fired upon by the limited anti-aircraft defences. On 1 January 1941 an Aer Lingus plane strayed into the Prohibited Area where the pilot was given a pyrotechnical New Year's greeting as he was smartly ushered back to the airport by rapid fire from guns in the Phoenix Park. Once or twice, the stately DH86 found itself in the middle of a tangle between the RAF and the Luftwaffe but the warplanes ignored it, being, in the words of one of its pilots, 'too slow to get out of its own way'!

The DC3, flagship of the fleet, was highly utilised, but its career was brought up short when a landing misjudgement at Liverpool caused severe damage to the aircraft, though thankfully none to its occupants. The wrecked aircraft was meticulously dismantled by Aer Lingus mechanics and shipped home for repair. It stayed out of service for four months, luckily when the company was least busy.

Another mishap to the small fleet occurred when one of the DH86s was on a test flight and suffered control problems. With limited control, the pilot got it down onto Portmarnock Strand where it nosed over. Luckily the tide, which would have engulfed the aircraft (shades of *Faith in Australia!*), was not particularly high and just lapped at its wheels. Technicians wheeled the aircraft to higher ground and, in appalling conditions caused by high winds driving stinging sand, repaired the damaged nose and other components, allowing a pilot to hop back to base.

The difficult conditions under which these aircraft were retrieved highlighted the excellent technical background that the airline had inherited from the Air Corps. Since its beginning, the Corps had difficulty in attracting suitably qualified staff, even when the catchment area included the UK. This was overcome in 1936 by a Boy Apprentice Scheme. Initially, the teenagers underwent eight weeks of military drill before embarking on comprehensive training, which included a general education subjects such as English, Irish, history and civics. In their second year, the boys were given more advanced training and earmarked for specific trades. In their third year they were transferred to the normal workshops and hangars to continue their training which was considered complete at the end of five years. A further intake started in 1937 and the scheme continues to the present day. The Apprentice School became a recognised training establishment for City and Guilds qualifications, and its success is evident in the subsequent careers of its pupils. Many went to Aer Lingus, others became pilots and, in time, senior airline captains but the majority continued their careers in civil aviation and rose to senior technical positions worldwide.

The senior ground technician, who had come over from the Air Corps to Aer Lingus at its foundation, always chose Air Corps men when a vacancy had to be filled. He had many good reasons – unlike other air forces, the Air Corps was of such small proportions that not alone did each man know his own trade, but was also something of an all-rounder. They were Army types, well used to being 'on parade' punctually and working all hours required for no extra remuneration.

Despite wartime difficulties, the airline continued to operate and by the end of 1943 the UK terminus was back at Liverpool again. But six months later this service was suspended, as the Allies girded themselves

ro engineers at work on a DC3.

We may not work in marble
(That would not pay a meal)
We may not skimp nor garble
Who needs must dream in steel:
Yet – given toil unshrinking –
Two ways the triumph comes;
The way of tireless thinking:
The rule of artful thumbs.

for the 'second front' – the Normandy landings. They feared that plans could be exposed to the Axis legations in Dublin, because of the travellers who constantly crossed the Irish Sea, even though the Irish authorities had clamped down on any possible espionage. The US minister of the time, David Gray, was behind a plan to get rid of these legations, despite their harmless nature. In collaboration with his British opposite number, he caused a 'diplomatic' note, bordering on an ultimatum, to be sent to the government. This approach, which became known as the 'American Note', was rejected. Like the various invasion scares of the previous four years, it brought the Army to the 'stand to'. In the end, the anti-Irish propaganda and news coverage rebounded on its perpetrators. The country had been seen to face up to the bullies, and the bullies stood down and subsequently blamed each other for the whole affair. The only Allied representative to come out well from this sorry episode was the Canadian representative – his government had, in the early days of the War, declared that, as a dominion, Canada had a right to remain neutral. When Churchill had requested that Canada associate itself with the ultimatum, he was turned down!

In the meantime, Aer Lingus was grounded for five months. But, with the end of the European War in May 1945, it returned to its task with renewed vigour and an unlimited worldwide market to exploit.

When Aer Lingus acquired five war-surplus C-47s for conversion to passenger-standard DC3s, the new purchases bore typical US Air Force 'nose art'. Seen here is one of the least daring examples, soon to be replaced by the shamrock and a saint's name!

Peace at last! Aer Lingus looked to the future and set about an expansion programme. An experienced pilot, Captain Ray Wells, was seconded from Trans World Airlines, which was happy to oblige a small company that it and the aviation world at large regarded as a mere 'feeder line'. Additional pilots came from the Air Corps and from the wartime RAF – though some of the logbooks of the latter were decidedly suspect. When it came to choosing cabin staff, Irish girls had no rivals! Ray Wells insisted that the girls show off their legs to the Board – and many a conservative eye was seen to sparkle! These hostesses went on to become a vital factor in 'the Friendly Airline', as it could be legitimately called in its early years.

The American captain was just as successful in getting the best possible aircraft fleet together: there were large numbers of surplus military DC3s (called the C-47), which had been the Allies' wartime aerial workhorse. Aer Lingus wanted half-a-dozen of these, but found itself at the end, instead of at the front, of a large queue of prospective customers – apparently through some clerical error. Ray Wells was not to be palmed off, and from an aerodrome in Germany he phoned the General Manager saying: 'I can get you six of the best, nil hours on the clock, for around £6,000 apiece.' In the event, the US authorities released five, and concurrently more were acquired, including a pair of excellent aircraft from the original Santa Monica Douglas production line. A standardisation programme to bring the aircraft up to passenger comfort standards by Aer Lingus and Scottish Aviation at Prestwick resulted in a fleet of ten DC3s and the two existing DH86s.

At the same time, the company looked to the Irish universities for new blood for its management trainee scheme. One of these was Garret Fitzgerald, later to be Taoiseach. Fitzgerald was just one of the team whose dedication to the airline, the State and the country at large saw Aer Lingus through its early, turbulent years. Despite the ups and downs of business life on the ground, which could be as violent as 'clear air turbulence' in the skies, Aer Lingus flew on steadily, and diversified into other endeavours – though not all of these were successful.

XIV

AER RIANTA:
MASTER OF MANY TRADES

WHEN AER RIANTA WAS ESTABLISHED by the government in 1937 as the *post facto* parent and guardian of Aer Lingus (and later Aerlínte Eireann) it was tasked with a general concern for all aviation matters. The new holding company was, in fact, the Irish aviation authority. Its overall stewardship of the operating companies meant that it shared their ups and downs in the cut-throat world of international civil aviation.

After World War II, Aer Rianta decided to complement Aer Lingus's scheduled services with its own mini airline, which would provide charter services operated by two Airspeed Consuls and a pair of delightful blue twin-tailed Miles Gemini air taxis. After a couple of years,

'The noisiest, most uncomfortable, roughest old crate we ever operated' was an Aer Lingus pilot's view of the Bristol Wayfarer – basically a freighter, though it did carry passengers. Here, a Wayfarer is being loaded at Dublin Airport with Irish exports, while its tanks are refuelled by Shell.

the tiny fleet was absorbed by Aer Lingus, which used the two Consuls for crew training and occasional charter work, leaving Aer Rianta with the management of Dublin Airport. Its philosophy was: 'to operate safe, efficient, customer-focused and environmentally responsible airport services on a commercial basis and competitive international rates', which it put into practice at Dublin, and subsequently at Shannon and Cork airports.

The success of Aer Rianta, now a semi-state company and on the threshold of public ownership, goes back to its early years. Personnel capable of putting its basic philosophy into practice came from the Civil Service, Aer Lingus and the Shannon Sales and Catering Organisation. Confounding the expected outcome, this assembly did not become a bureaucracy. What eventually evolved was a 'flat organisation', where decisions did not filter down through layers of management, but one where everyone, from top to bottom, was involved in decisions. This strong 'human resource' focus has worked, and worked well.

Dublin Airport was the focus of much management attention in 1969, as Aer Rianta was faced with the imminent arrival of the Boeing 747 jumbo jet. The sheer size of the B747 produced huge problems for all airports and Dublin was not alone in having to undertake a massive building programme to accommodate its passenger load. Build it did, however, and in 1972 a new passenger terminal opened its doors to the jumbo jet age. So, thirty odd years after the first Aer Lingus flight took off from the newly-built airport, on 19 January 1940, Dublin Airport was catering for 1.7 million passengers. Today, a further thirty years later, this figure has risen to 14 million.

Shannon, too, demonstrated agile marketing. When, back in 1937, Rineanna was advancing to its final status as Shannon International Airport, the government had proposed that Aer Rianta should be placed in control of this project. However, it was to be thirty years before the company took over the

Early days at Shannon:
Ground hostesses escort passengers to their aircraft – reassuring words were always helpful in the days before customers became blasé!

running of this major aerial crossroads. Some years ago, when Shannon was no longer an obligatory transatlantic stop, the prophets of doom expected it to wither and die. They were confounded! In 1999, over 2 million passengers passed through the airport, representing almost twenty percent growth over the previous year – at a site where once the marsh birds reigned supreme!

During the Cold War, when the Soviet Union would rarely get involved with anyone outside its sphere of influence, the Russians were prevailed upon to use Shannon Airport in a unique barter arrangement, whereby Aer Rianta traded its charges for oil! A fuel farm was set up that enabled Russian tankers to offload oil supplies that were then pumped, virtually directly, to Aeroflot planes on the Shannonside tarmac. This arrangement worked so well that a joint company, called Aerofirst, led to the Irish company managing and overseeing the first Duty Free shop in Moscow. In that field Aer Rianta had pioneering experience going back to the first ever Duty Free shop in the world, which opened its shutters at Shannon in 1947. Despite the meddling of Brussels bureaucrats in abolishing intra-EU Duty Free sales in 1999, Aer Rianta has accelerated its overseas development in Duty Free retailing through its subsidiary, Aer Rianta International (ARI). It is now one of the largest Duty

Mrs Bertie Ring in the World's first Duty Free shop, opened by Aer Rianta in Shannon. A wide range of high-quality goods, mostly Irish, were available at rock-bottom prices.

Free retailers in the world, with a managed turnover of US$300 million and interests in fourteen countries worldwide.

The last year of the old millennium was a good one for Aer Rianta, with passenger numbers growing to 16.5 million at the three airports. Despite losing revenue from the abolition of Duty Free sales, the income from landing fees and passenger service charges amounted to £52 million, or just £3.11 per passenger, the lowest of any comparable airport operator. During the year in question, thirty-four airlines operated scheduled services to eighty different destinations emanating from the three Irish airports, but a rising number of air travellers brings with it the need for extra facilities. At Shannon a new terminal building and road system were put in place, while at Cork a new pier, incorporating 'air-bridges', will enable that airport to take 3.5 million passengers per year.

Rolling out the turf as Dublin Airport is being completed in 1939.
Concrete runways would not be laid until the end of World War II.

Air travel is expected to double in the next decade, and Aer Rianta is planning on building for the future. Dublin Airport's brand new terminal extension will open in autumn 2000, and plans for a new parallel runway are well advanced. However, there is a downside to airport expansion: the conflict between economics and aesthetics. When Dublin Airport was opened in January 1940, the 'jewel in the crown' was the terminal building designed by Desmond Fitzgerald (Garret's older brother), which deservedly won for him the RIAI Triennial Gold Medal in 1943. In its time it dramatically symbolised the romance of air travel, which, of course, has moved on from an era when one walked or was bussed out to the aircraft. Nowadays it is more usual to walk dry-shod along the air-bridges direct into the aircraft, and this has a major impact on airport architecture. At Dublin Airport, the busiest of Aer Rianta's operations, lengthy piers have sprouted – the latest one is 200 metres long – and overwhelmed Fitzgerald's masterpiece. It is surely difficult to serve both the god of aesthetics and the mammon of economics!

Desmond Fitzgerald's architectural gem can be seen in all its brilliance, just as traffic was building up after World War II A couple of redoubtable DC3s are on the apron and, between them, one of the unloved Bristol Wayfarers.

Aer Rianta's international operations are handled by Aer Rianta International (ARI), a venture that promotes a variety of commercial activities and services, which a relatively small home market with a population of 3.6 million could not sustain. Brains and experience were exported! So it is that from its small beginnings in 1937 as a guardian of the infant Aer Lingus, Aer Rianta now provides the master planning, design, marketing and financial management of major airports on a global scale. In 1999 ARI's overseas activities added more than £7 million to the overall corporate profit, and today the turnover of ARI exceeds that of its parent company.

The all-over endeavours of Aer Rianta, with ARI, have not gone unnoticed in world industry to the extent that it received the Eagle Award from the International Air Transport Association (IATA), the world body that represents over 300 airlines globally. The Eagle Award is a recognition of Aer Rianta's excellent contribution to generating aviation business, and its setting of a benchmark to which other airports could aspire.

On the gentler side, and away from the hard-nosed world of airport management, Aer Rianta has been responsible for making the country air-minded. Its 'air education programme' has reached into schools and colleges, and the company has sponsored many of the Air Spectaculars that enliven the summer months for huge crowds at Baldonnel, Leopardstown and Cork. Then there is the Heritage Programme, which reflects Irish identity and accomplishments, both at home and abroad, and celebrates special occasions throughout the year: a Carnival of Lights at Christmas, concert recitals and an arts festival are just some of the jollifications. More seriously, Aer Rianta has a very positive attitude towards the environment and has been behind many landscaping projects. These are merely the outward signs of the company's fulfilment of its goal to be 'the best organisation in the world in the field of managing airports and associated commercial activities'. Aer Rianta may not be a 'jack-of-all-aviation-trades', but it is certainly a master of many, and though its feet are firmly planted on the tarmac of many an airport, its eyes are firmly focussed on the future.

XV

WATCH ON THE WEST

IN 1936, THE CHIEF OF STAFF OF THE DEFENCE FORCES RECOMMENDED that a three-year purchasing programme for equipment be initiated; he stressed that the Air Corps and anti-aircraft defence should be given special attention. Nothing happened, and it took the Munich Crisis, Hitler's takeover of Sudetenland in 1938, to provide the spur. But it was a case of too little, too late. Military supplies began to trickle through in 1939 but the chief supplier, Great Britain, had to look to its own needs and cancelled some of the Irish orders. When the first nine months of the war, the 'sitzkrieg', turned into 'blitz-krieg' in the summer of 1940, and invasion from both sides was a possibility, Irish defences were pathetically inadequate. Despite the rush to join the expanded Army, suitable equipment was at a premium. The Air Corps had a mere handful of reasonably modern service aircraft and trainers. Now panicking, the government asked Britain to

Three lieutenants peruse a map during pre-flight briefing. The silhouettes on the wall show the various types of vessels that might be encountered.

train a hundred fighter pilots and said that it would gladly purchase the warplanes for them. But there was no chance of this happening!

Despite minimal resources, the Air Corps was doing its best. A couple of days before the outbreak of war, a newly formed cadre of the

'Winding up a Walrus!' The inertial starter required elbow grease.

First Coastal Patrol Squadron commenced operations from Rineanna in County Clare. Its mission was to patrol the west and southwest coasts, up as far as Donegal and down to Wexford. The unit had four Avro Ansons, a pair of Walrus amphibians, and a runabout Cadet biplane. Its personnel totalled seventy-four of all ranks, the airmen being billeted in huts similar to those occupied by the construction workers working on the concrete runways. For over a year there was no hangar; aircraft had to be picketed in the open and flown back to Baldonnel for twenty-hour inspections until a large three-bay hangar was erected.

A Walrus amphibian heads outwards over the Shannon estuary on a patrol.

Just before Christmas 1939 the unit lost its first aircraft. A patrolling Anson with a three-man crew suffered a double engine failure and had to put down in Galway Bay, where it began to sink rapidly. Luckily the Anson had a rubber dinghy which inflated on contact with water, and the three airmen scrambled into this. They were soon overwhelmed by high waves, but on the Galway shore their plight was observed by local fishermen and a currach was rowed out. The three were rescued, but not before the wireless operator coolly returned to his virtually submerged aircraft to retrieve his watch! Thus began the attrition on aircraft which gradually reduced the original nine Ansons in use at the beginning of the war. Seven other Ansons had been ordered pre-war but, though these were finished in Irish colours and ready for imminent delivery, as soon as war broke out the RAF commandeered them for its own use.

Patrols of approximately three hours' duration, at a distance of up to ten miles from the coastline, were flown regularly, often in the most appalling weather conditions. The airmen kept their base informed of all seaward activity and, in particular, of weather conditions. Despite well-publicised rumours in the

Above: Bombing-up an Anson. The coastal patrol aircraft were usually armed.

Right: The crew board an Avro Anson before setting off from their Rineanna base.

British and US press, no German U-boat ever received shelter or succour in Irish territorial waters. The patrolling pilots were meticulous: 'We got so used to our beat, that even if a small rowing boat was not where it should be, we would investigate!' The patrols often spotted marauding and meteorological German Condors taking shortcuts across the southwest. To counter these intrusions, Rineanna would broadcast their courses on a very wide wireless band, and RAF aircraft would soon be observed in hot pursuit.

Early on, David Gray, the American minister in Dublin, requested that the coastal patrols be terminated short of the Wexford 'slobs' – apparently, the aircraft were disturbing the wild geese feeding there, thus spoiling his wildfowling activities! This frivolous complaint got the reply it deserved. Diligent observance of intruding belligerent aircraft and naval vessels, drifting mines and survivors of sunken ships were vastly more important. Gray had received ministerial rank because he was related by marriage to President Roosevelt. He would send Roosevelt the results of spiritualistic seances which proved that Ireland was in league with the Germans!

The patrols continued, but attrition of the heavily-utilised aircraft and their demands on scarce fuel supplies caused a gradual reduction in their frequency. The British were upset but did not provide the necessary aircraft. The Irish authorities considered that seaward surveillance was now amply covered by a chain of over eighty look-out posts set

Above: A downed RAF Lockheed Hudson that the Air Corps repaired and used on the coastal patrol. The Hudson was derived from the Lockheed 14, a pair of which Aer Lingus sold just when the under-equipped Corps needed them!

Left: Deserted now, this is one of the ninety-odd look-out posts from where coast-watchers liaised with their airborne comrades.

up on headlands and promontories and manned by what the 'fighting army' dubbed 'The Saygulls'. Midway through the war, air patrols were only resumed on demand, in cases where the coastwatchers called up Rineanna to ask for a closer look at a specific situation.

Though the pilots took their official duties seriously, the younger ones got their kicks by some crazy flying, away from the eyes of authority. One of these coastal guardians confessed:

> We had one Cadet biplane as a sort of a runabout. It was a very forgiving machine, even for the ham-fisted; loosely-rigged, you could get away with a lot of silly flying in it. And just as well, for in a foolhardy competition which myself and two pals dreamt up we spun the Cadet from 2,000 feet to see who could get in the most turns before pulling out. The winner on one occasion had a very near miss, hitting the top branches of trees on recovery!

'Rounding the F–ing Fastnet.' A study of an Anson and a description by a hard-pressed coastal patrol pilot!

For the pilot quoted above, and others, life at Rineanna, a damp and isolated spot, could be very boring. While under open arrest for some small misdemeanour, he took off in one of the Walrus amphibians with the intention of reaching Cherbourg in occupied France, there to join the Luftwaffe, the German Air Force. However, over Cornwall, he was forced down by RAF fighters and returned to Ireland to face the music. Having paid for his sins, he joined the RAF and had a distinguished record in action, followed by an equally successful career as a businessman.

The Rineanna crews were always on hand to render aid to both sides in the conflict. On one occasion a signal was received from the RAF saying that one of its Sunderland flying boats was in trouble and had lost a float, rendering it unable to alight on the sea. It was a wild January night, and the Air Corps proposed to fly out to meet it and guide it into a landing on the grass at Rineanna. Unfortunately, as the patrol aircraft was about to take off on its mission of mercy, a further signal confirmed that the large flying boat had, in fact, landed off the Clare coast. Without its float, it had turned turtle, with the loss of nine of its eleven crew.

By 1943, as the Allies began to win the Battle of the Atlantic, what remained of the Coastal Patrol unit returned to Baldonnel and was disbanded, its place being taken by a Hurricane fighter squadron. The Coastal Patrol Squadron had shown from the beginning of the conflict that Ireland was serious about its neutrality – however one-sided that neutrality may have been!

At the request of the US Army Air Force, the headlands where the look-out posts were positioned were given thirty-foot numbers, together with the word 'EIRE' to assist aircrews. This view of a Wexford navigational signpost was taken from one of the coastal patrol aircraft.

The Coastal Patrol Squadron based at Rineanna was replaced by the First Fighter Squadron, seen here in 1944, preparing for an inspection by the officer commanding the First 'Thunderbolt' Division.

XVI

THE WRACK OF WAR

HEAVY, BLUSTERY RAIN covered the east coast of Ireland on the morning that Britain declared war on Germany. Out of the murk an RAF twin-engined Lerwick flying boat emerged and alighted off Dún Laoghaire. Undisturbed by the authorities, it left shortly afterwards, just as a larger four-engined Sunderland flying boat was also driven down at Skerries by the appalling weather. The crew was unaware that war had broken out, and the pilot was given an excellent lunch in an Army mess before the Lerwick returned to render assistance. Both crews were treated as 'distressed mariners' and allowed to depart, enhanced by a gift of Guinness to the Sunderland's pilot. This may have been because of the name he bore – he was Squadron Leader Michael Collins! Eighteen other RAF intruders were allowed to depart before a somewhat stricter policy of neutrality was adopted.

There was not much 'belligerent' aerial activity over Ireland until the fall of France in 1940, but by the end of the war well over 200 crashes and forced landings had occurred in Ireland and her territorial waters. Though the records are imprecise, it appears that there were a minimum of 120 RAF and Canadian, forty United States Air Force and twenty-five German warplanes which crashed on land or went down 'in the drink'. There were many 'touch-and-go' occurrences like the Skerries incident, many lucky escapes from disaster and, sadly, many accidents which caused complete destruction of aircraft and loss of all crew.

In those now far-off years many people in Ireland had never seen an aeroplane, and certainly few had seen one at close quarters. The various air displays of the 1930s had been staged in cities and major towns. Flights by private pilots and the tiny Irish Air Corps would rarely disturb the rural calm. Imagine then the spectacle of a huge, heavily-armed, four-engined war machine coming like a bolt out of the blue, literally, and performing dramatic manoeuvres to effect a landing. Such dramas were the talk of the towns and the villages for years. In fact, many localities have tangibly marked these awesome events in memorials such as wall plaques or crosses, commemorating triumphs or tragedies.

When a crash occurred, someone had to pick up the pieces. Though the Air Corps had insufficient aircraft to perform its designated duties, its technicians were still kept busy, sometimes fielding three salvage teams to dismantle, repair or demolish downed aircraft. These teams undertook their tasks with a high degree of professionalism, despite the initial scepticism of the Allied authorities. Often they operated in impossible terrain, such as inaccessible mountains and bogs, and in the worst extremes of weather. The Allies were anxious to retrieve valuable aircraft intact or at least all the weaponry and instrumentation that could be salvaged. From the Irish standpoint, it was important to ensure that wrecks would not pose a hazard to life and limb, and that arms and ammunition would not get into the wrong hands. Generally, there

Lieutenant Georg Fleischmann, attached to the Luftwaffe ...

... who took this picture from his Heinkel III over the Bristol Channel on 1 April 1941, as a trio from the Luftwaffe base at Tours attacked shipping, scoring a direct hit on the tanker San Lonredo.

The Heinkel was attacked by British Spitfires from Pembry aerodrome, which damaged its starboard engine and undercarraige, and wounded the pilot, Lieutenant Heinz Grau. The Heinkel made a wheels-up landing in County Waterford, being stopped by an earthen bank. The crew used a destructive device, which failed to operate, and then used a machine-gun to wreck the instruments.

was very little left of German aircraft, because planes and equipment were destroyed by surviving aircrew who would even, on occasion, hold off local forces until destruction was complete.

Lieutenant Jim Teague of the Irish Air Corps (on the right), examining the crashed Heinkel III. An Air Corps salvage team dismantled it and brought it to Baldonnel.

Surviving German and British airmen were interned at the Curragh Camp, Ireland's largest military base. For both sides it was a benevolent internment – each day, the internees could sign out on parole and roam anywhere within a large area of County Kildare. They had their own football teams which frequently played against their Irish Army custodians, and some became members of nearby golf and tennis clubs – a number of RAF officers even rode to hounds! Once a month, and later once a week, the internees were allowed

Lieutenant Fleischmann and his crew ended up with other Luftwaffe and Kriegsmarine (Navy) personnel as internees at the Curragh, where Herr Fleischmann took this photograph. After the war, he returned to work in Ireland, before relocating to Canada.

a day visit to Dublin, and eventually several internees pursued studies at Trinity College, Dublin, and University College, Dublin. Ireland's 'special relationship' with the US decreed that no Americans were confined. The RAF internees made several escape bids in the hope of getting across the Border. Some succeeded, and were again posted to operational flying, only to die in combat. Gradually, RAF men were repatriated and by the spring of 1944 all had been released. In the case of the Luftwaffe and Kriegsmarine sailors, there was no way they could be returned to the Fatherland so they had to sit out the war. At the end of hostilities, as international law required, they were handed over as prisoners of war to the British. Many subsequently returned to marry Irish women and settle down as free men, far from the chaos which had greeted them at home when they were repatriated.

A US Air Force Liberator, down in a mountain bog in County Mayo after a long Atlantic flight, where an Irish Air Corps salvage team dismantled it.

A typical chore for an Irish salvage team resulted from the forced landing of a US Liberator on its last gasp after a transatlantic flight, when it put down in a mountainous bog in County Mayo. Its bomb aimer, who returned to the scene years later, describes what happened:

We were flying over from Newfoundland and had been fourteen hours in the air when we ran into heavy squalls. All our navigation instruments went out, and we were very low on fuel. I was in my position in the nose and our pilots asked me to keep an eye out for any break in the cloud. I remember seeing green grass, which looked as if we were over Ireland – our destination, in fact, was a Northern Ireland base. When I spotted a large hole in the cloudbank our chief pilot said: 'I'm going down to see if I can get a place to land.' There were cattle in the fields, showing, we thought, that the ground was solid and we scared them off by low flying. Then we touched down on what turned out to be a soft peat bog which tore off the landing gear. The bomb bay doors burst open and scooped the peat into the plane. This slowed down the forward movement and acted as a cushion. I remember one engine was smoking, and there was a bit of fire, but rain came and put it out. Miraculously all eleven of us got out with very minor scratches. The first person to come up was an Irish policeman, who told us not to worry because we would be at our Northern Irish base within seventy-two hours. We were entertained lavishly in Foxford and we hardly slept at all for those seventy-two hours!

The tail guns of the Liberator that came to rest in a Mayo bog.

A kindly Garda presents a young gossoon with a souvenir from the USAAF Liberator.

Commandant Jack Ryan, an Air Corps engineer, now takes up the story:

> The Yanks only wanted the engines, instruments and all removable gear, so the large airframe was left. I got many quiet approaches from local scrap merchants for exclusive salvage rights. However, when we had recovered all the required equipment, I posted notices saying that the plane would be blown up at 1500 hours on the day our work would be completed. As that hour approached, about six lorries and a dozen farm-carts lined up behind our cordon. When the demolition charge exploded, off went everyone in a mad dash; we even had to fight hard to get our own equipment clear! A pair of local electricians were soon madly stripping switches and wires out of one section of the fuselage, when six burly fellows, employed by a rival scrap man, picked up the whole section with these two inside, tossed it onto their lorry, and drove off. Some little way up the boreen, the scrap merchants kicked out the electricians and proceeded on their way! There were fist-fights and threats all over the place, but eventually everyone simmered down – except the Gardaí. During the previous fortnight they had been running in locals for even taking small souvenirs from the crash, but now people were making off with lorry loads of the stuff!

The same engineer officer recalled an earlier incident up in the Border area, where aircraft from Northern Ireland bases were constantly coming down:

On the Mayo operation, we had low-loaders which had formed part of a consignment of salvage equipment, including five-ton electric cranes, jacks and trestles, supplied by the British in 1942. Properly equipped at last, we reckoned we could move anything out of anywhere. In contrast, before this equipment was acquired, my team was attempting to salvage an RAF Hampden bomber which had come to rest in a potato field. We floundered and struggled in a sea of mud, but eventually got most of the aircraft out on hand-carts. However, the large radial engines, each weighing a ton, could not be shifted despite the all-out efforts of all ten sturdy men. I remembered an old hypnotic trick and told the lads to place their hands on the propeller shaft, one on top of the other in rotation, and to press down hard. I then ordered 'break and lift', whereupon all hands heaved together. This time, the engine virtually floated up onto the cart, and my men looked at each other in wild surmise and marvelled at the power of auto-suggestion!

The Luftwaffe dead lie in the German War Cemetery in the Wicklow mountains, in a place of rare beauty and serenity. A plinth there is inscribed in Irish, German and English:

> It was for me to die under an Irish sky,
> There finding berth in good Irish earth;
> What I dreamed and planned bound me to my Fatherland,
> But war sent me to sleep in Glencree.
> Passion and pain were my loss, my gain;
> Pray, as you pass, to make good my loss.

The new US President, Harry S Truman, who had succeeded Franklin D Roosevelt before the war ended, decreed that 1 August 1945 be celebrated as 'Army Air Forces Day'. The US minister was host at a dinner in Dublin, and though he had been strongly antagonistic to Irish neutrality, he had the good grace to express the gratitude of the American people to the Irish forces for their 'unfailing kindness and tenderness to "shipwrecked" airmen, whose wounded had been cared for with solicitude and whose dead had been buried with the greatest respect'. Lieutenant General Dan McKenna responded: 'I am glad that Ireland tended to the downed airmen in the way they deserved, and rendered to all airmen, of all nationalities, the assistance and honour which was their due.'

XVII

SHANNONSIDE I – RINEANNA

THE DEVELOPMENT OF A TRANSATLANTIC AIR SERVICE to link the Old and New Worlds began in 1935 at a conference held in Ottawa, where an agreement was signed between Britain, Canada, Newfoundland (which was then independent of its neighbour) and the Irish Free State. Agreement was reached on passenger travel and postal services, and the participants chose an operating company – it was Imperial Airways, the forerunner of the British Overseas Airways Corporation (BOAC). The flying boat was the chosen vehicle, because land planes of sufficient passenger payload and range had not yet been developed, nor were there any suitable airfields for them. Ireland was to provide an Eastern Air Terminal, which would offer combined land and sea facilities. When ready, the United States would be granted landing facilities at this terminal and the then mighty Pan-American Airways would operate reciprocal services. A further arrangement, which was never completed, would grant the Irish Government shares in an overall financial plan.

To select a suitable Eastern Terminal site, surveys were carried out at Valencia Island, Kenmare and Tralee Bays in Kerry; Cork's great harbour; Galway Bay; and Lough Ree in the middle and Lough Derg on the lower reaches of the River Shannon. Germany, though not involved in the transatlantic agreement, was interested in Galway. In September 1936 Deutsche Lufthansa had flown two of its Dornier flying boats from Berlin to New York, refuelling in the mid-Atlantic from a tanker. The intention was to test the suitability of Galway Bay as a stop-off point on the return flight. Galway Corporation had set up the Irish Trans-Atlantic Corporation and appointed the pioneering pilot, Sir Alan Cobham, as technical consultant. He picked out a site at Rusheen Point, three miles out from the city. However, the Irish government, quoting the Ottawa agreement, turned down the German plan, and Lufthansa re-routed the return flight via the Azores. Hitler was now in power and it is very likely that this was a factor in the decision.

By the end of 1935, the survey party had finally settled on the spot where the Fergus River in Clare flows into the mighty Shannon, at Rineanna. This name, a corruption of the Irish for 'promontory of the birds', covered a large marshy area which could be drained. The Air Corps carried out a photo survey and the work of clearance and drainage was well advanced when Colonel Charles Lindberg took a look both from the air and on foot and gave his enthusiastic approval. Concurrently, construction of a main communications station at nearby Ballygirreen commenced and it was formally announced that the complete facility would be known as 'Kilconry International Airport' as this is the townland in which Rineanna is situated.

Above: Corporal Charlie Rooney, photographer, and Lieutenant Johnny Moynihan, pilot, who conducted the aerial photo-survey of Rineanna in a Vickers Vespa in 1935.

Right: Pinning the mosaic of photographs which resulted.

Cork and Galway noses were out of joint, but there were good reasons for the Shannonside choice. Ireland's greatest river could easily be spotted from the air. It was navigable for forty miles from Loop Head up to Limerick City, which would facilitate water-borne traffic, and it was sheltered from the Atlantic winds and heavy seas. Botwood, a seaport in Newfoundland, was chosen as the Western Terminal in North America. It would conveniently connect with New Brunswick, which would provide a refuelling point en route to New York City. The tranquillity of all three points, in Canada, Newfoundland and Ireland, would soon be shattered by the roar of mighty engines but, on the positive side, the local economies would benefit.

Colonel Charles Lindberg, 'The Lone Eagle', approves the Rineanna site on behalf of PanAm.

While Rineanna was a-building, Foynes, a seaport ten miles upriver, would provide an interim flying boat base. Rineanna, when complete, would incorporate a basin to take fifteen large flying boats. However, fortunately for Foynes, construction of the basin was held up by the lack of a suitable dredger vessel, due to wartime shortages. In the long term this was just as well, because the flying boat was soon to disappear – though, of course, the basin might have made an ideal yacht marina. As well as being a base for the Air Corps Coastal Patrol Unit, Rineanna became the link for passengers and cargo between the flying boats and landplanes. These were bussed between the two bases. DC3s, civilianised Whitleys and Hudsons and the beautiful four-engined De Havilland Albatross airliners operated connecting flights to and from the UK. In July 1943, one of the latter required Air Corps repairs to its stressed plywood skin while another, coming in to land with half the BOAC board as passengers, had a more serious structural failure, when the glue holding the wooden frame together failed. Army medics were quickly on the scene but luckily there were no serious injuries. Incidentally, the anti-aircraft guns guarding Rineanna would follow the civil aircraft in and out by way of practice – which, understandably, pilots found somewhat disconcerting!

The elegant DH Albatross 'Fortuna' at Croydon, London's pre-war airport.
The aircraft's wood and glue construction, however, failed as it came in to land at Rineanna in 1943.
Mercifully, the only casualty was a Clare goat who got in the way!

In addition to military and civil activities, planes from the US Army Air Force were frequent, if unintentional, visitors to Rineanna. They were on ferry flights, as distinct from offensive operations, and all were allowed to continue their journeys. But a precedent had to be established. When, in July 1942, the first of the US aircraft, en route from the United States via Labrador and Iceland to Scotland, landed, the station fire engine blocked any attempt at a getaway. As the occupants deplaned they were arrested and their weapons confiscated. There were fifteen aircrew on board, destined for the build-up of the Eighth Air Force in Britain. De Valera himself authorised the release of aircraft and crew, who were provided with an exceedingly good night's entertainment. Next morning, a local test flight was made with Air Corps pilots on board, and then the Americans went off to war.

There were nine more US Army Air Force visitors, and on occasion Rineanna looked like a US base. At one time, two of these aircraft, which arrived soon after each other, waited over during repairs and refuelling. The Air Corps men who were attending to these operations were surprised that there was no fraternisation between the two aircrews. In an unmistakable southern drawl one pilot explained that the other aircraft was manned by 'damned Yankees', and his men wanted nothing to do with them. The bitterness of the American Civil War lasted even longer than the effects of the Irish Civil War!

The only casualty occurred when an Air Corps mechanic was repairing an engine on a Flying Fortress. The engine exploded in flames, blowing the unfortunate man off the high wing to a hard landing on the concrete, which broke his leg. There was another explosion when the crew of a Liberator demolished its radio equipment and secret Norden bombsight. Nevertheless, no offence was taken by the Air Corps. They refuelled it and waved it off on its way to Northern Ireland.

The American airmen throughout this time thoroughly enjoyed their stopovers and said that they hoped they would get as good a reception from the English! At the end of hostilities, when Rineanna became Shannon International Airport, American civil aircraft arrived in droves, laden with Irish-American tourists.

Early days at Shannon International Airport: a mobile navigation aid guides a BOAC Lockheed Constellation to a safe arrival. This particular aircraft had, in fact, been St Patrick, the first Aerlínte aircraft, which, with four others, had to be disposed of when the Atlantic project was cancelled.

XVIII

SHANNONSIDE II – FOYNES

THE FLYING BOAT CAME INTO ITS OWN IN CIVIL AVIATION when it made the North Atlantic a viable modern airway. As we have seen, Rineanna was designated as the Eastern Air Terminal, and the small town of Foynes, slightly upriver on the opposite bank, provided *pro tem* facilities until the planned flying boat basin could be completed. No one knew that the flying boat and the basin at Rineanna would become completely redundant in the short space of a decade. By the mid-nineteenth century, Foynes had developed into a considerable port with fuel storage facilities and a railway link. Its harbour was sheltered by an island and the intervening channel was an ideal haven for the airborne boats.

The Mayo Composite 'Pick-a-Back' aircraft. The flying boat would assist the seaplane, heavily laden with mail, until they seperated over the ocean and the mailplane's own efforts would bring it on to New York.

A large pier, added in 1936 in anticipation of its role, could accommodate tankers loaded with aviation spirit and other petroleum products. Meteorological and communications services were established in the Monteagle Arms Hotel, which had been adapted as a control centre. Based here, meteorologists could work out the weather conditions that aircraft were likely to encounter and calculate the average time it would take to 'leap the water-jump' at a given speed on any day. Radio officers maintained the ground-to-air links with their opposite numbers in the aircraft. In those days, and indeed until the mid-1950s, communications meant radio-telegraphy tapped out on a Morse code key. In addition to the main communications station at Ballygirreen, near Rineanna, there was a direction-finding station at the topmost point on Foynes Island, to assist incoming craft when the visibility was poor. Foynes control was responsible for radio communications as far as the mid-point of the Atlantic, the western sector being covered by the Gander station in Newfoundland.

Everything was ready for the first test flight when, at the end of February 1937, an Imperial Airways Empire flying boat named *Cambria* took off from the company's flying boat base at Southampton. It alighted on the Shannon, to the cheers of hundreds of onlookers. Next morning, its huge tanks filled with Shell spirit, *Cambria* taxied into mid-channel and within a minute was 'on the step' (that is, planing along on the sharp chine of its hull) and off the water. The flight covered 400 miles along the west coast and all systems worked perfectly, despite rain and hail which forced the flying boat down to 1,000 feet and occasionally reduced its speed to 200mph. The next day, the best landing areas on the river in terms of currents and tides were investigated. A green light was given for transatlantic operations and *Cambria* departed for its British base.

Appropriately for the Americans, on Sunday 4 July, as the Imperial Airways flying boat *Caledonian* was preparing to leave Foynes for North America, Pan American's *Clipper III* left New York. Twelve-and-a-half hours after leaving Botwood, *Clipper III* roared low out of a murky sky over Monument Hill at Foynes as a pair of Coastal Patrol Ansons circled above it in welcome. On shore, government ministers greeted the seven-man crew, whose most important member, according to the captain, was the steward. He had served an excellent mid-air meal – the hallmark of flying boat luxury and comfort, which to this day provokes nostalgia in those who were lucky enough to experience it!

Left: Foynes 1943. Bad weather keeps virtually all the flying boats on the water, but not the American Export Catalina, which somehow managed to get into the picture.

Right: The two components of the Mayo Composite pick-a-back experiment at Foynes.

Left: Father Rice, a local priest, sees off fifty pilgrims to Fatima on Aquila Airways's Sunderland. This airline operated from 1948 to 1958.

Thus, as sabres rattled in Europe, the air link joining two great continents was forged. Unfortunately, the science of flight was soon deployed in a savage war, during which Foynes became a vital factor in the Allied cause, despite Ireland's declared neutrality. Military and political VIPs and warlike stores passed through regularly. Security was stringent and the German legation was kept in the dark as far as possible as Allied traffic increased. The Irish authorities never capitalised on this benevolence in terms of political gain, nor in terms of practical matters, such as food and badly-needed military supplies. In fact, apart from landing fees, the country never profited from Shannonside's key strategic position during World War II.

Take-off from Foynes was usually at night, as this timing would permit a landing, a more critical operation, in daylight on the other side. In addition, the cloak of darkness lowered the risk of interception by the far-ranging Luftwaffe Condors, which Churchill dubbed 'The Scourge of the Atlantic'. Watching a night-time take-off at Foynes was an exhilarating sight! As take-off time approached, a launch would go out to place lighted buoys at 200-yard intervals to form a mile-long flare-path down the shipping lane. At the upwind end of this 'runway' the revving engines of the flying boat would produce a powerful roar. When all was set, the captain's voice would be heard through a loudspeaker in the control launch: 'Ready for take-off – release flare number one.' Up into the night sky a parachute flare would rocket and explode into vivid white light, turning night into day. The flying boat's heavy bulk generated a great wall of spray until, gathering speed all the while, it would 'unstick' from the water, which seemed reluctant to let it away. Another command to the control boat and a second rocket would shoot up to revive the dying light of the first and bathe both Clare and Limerick in its brilliance.

Initially, wind and currents would move the lighted buoys out of position and present a somewhat serpentine flare-path, but this was overcome when a fixed marker light system was installed. A landing in the early dawn was almost as dramatic, though perhaps not for the launch crew – weather could delay the estimated time of arrival by up to an hour, which could mean a chilly wait. The Ballygirreen radio station would pour out bearings to the incoming craft and parachute flares would supplement the dawn light like welcoming beacons for the weary travellers. Unfortunately, stringent wartime restrictions on photography prevented cameras from capturing these dramatic moments.

Servicing the engines as a flying boat lay at anchor could be difficult – its wings would often be wet and slippery and even though their leading edges could be folded down to form a working platform, a sudden sway of the aircraft could cause trouble; a dropped spanner was a lost spanner unless it was tied to a line which had a float. After a night upon the water, the hull and floats needed to be checked. A little seepage was acceptable, but should the wing floats have taken water, a hand-held bilge pump would be needed. As engines were closed down after a flight or a test run, the propellers were inched around by the starting motor so that they would not obstruct the launches.

One of Ireland's few seaplanes, a type of aircraft which could have been invaluable in World War II, in the view of General MJ Costello, lands on the Shannon river in 1940. This actual aeroplane, a Vickers-Supermarine Walrus, was salvaged from a wrecker's yard and completely restored, as the last of the 740 built.
(This type was nicknamed the 'Shagbat'.)

BOAC was steadily expanding its presence at Foynes. From a modest site at Boland's Meadow, a large complex grew with facilities for passengers, a clubhouse, a dozen dwellings – six of which were occupied by company employees and their families – and a restaurant. It became known as 'The Colony', and a very British one it was too! Some of the senior managers were extremely anti-Irish and one of them, whose bullying ways were not appreciated by either British or Irish staff, was particularly offensive. One of his most blasphemous outbursts caused the large number of Catholic employees to down tools until an apology was forthcoming. But generally British and Irish got on well together, even when The Colony celebrated VE Day. A senior BOAC manager felt that the display of the Irish Tricolour at the celebrations was 'completely out of place'. He became very agitated and demanded that the committee remove it, but instead his colleagues put up a second one! The Irish Army officer who controlled the base was seething, but advised everybody to ignore the manager's 'very bad taste'.

Foynes was heavily protected by the Army and Garda, but despite this BOAC initially pressed for armed guards on their aircraft. The Army commander imposed strict orders regarding the wearing of uniforms by Allied military personnel, who were using Foynes in increasing numbers. On one occasion, he ordered three British colonels to remove their rank markings and to carry their overcoats on their arms to disguise their tunics.

Security was strict at Foynes. Locals applying for jobs there were always vetted by the Garda to intercept would-be saboteurs, and, as seen here, a pistol-packing policeman guarded any flying boat not moored out in deep water.

Captain Jack Kelly-Rogers was one of the Imperial Airways pilots to span the transatlantic route on the proving flights in the Short Empire flying boats *Connemara* and *Caribou*. A Dublin man, he was extremely professional and did not suffer fools gladly, but he could take a joke. In the days before his company acquired its three Boeing 314 flying boats, it had been necessary to ferry the passengers and crew ashore by launch. This became redundant when the Boeing's 'sponsons' (stub wings at water level instead of wing floats) allowed disembarkation directly to the quayside. Once, in pre-Boeing days, the launch took Kelly-Rogers's crew and passengers ashore, but for some reason not the great man himself. The coxswain returned and, tongue-in-cheek, said, 'Captain, I was sure you could get ashore under your own steam!' Kelly-Rogers replied: 'Coxswain, I know my initials are JC, but walking on water is not one of my skills!' After the war, KR, as he was always known, became Technical Manager of Aer Lingus. He also found time in the 1970s to initiate an aviation museum, initially in Dublin Airport and then in nearby Castlemoate House, which, regrettably, has lapsed.

Tranquility at Foynes: this peaceful scene is soon to be shattered by the four powerful engines of the Boeing 314 when they burst into life. The aircraft's 'sponsons' may be seen just above water level.

Much to the annoyance of BOAC and PanAm, who considered that they had the transatlantic route sewn up, a newcomer appeared in the shape of American Export Airlines, a subsidiary of a major liner company. In the summer of 1939, its Catalina flying boat, *Transatlantic*, made three round trips across the Atlantic just when PanAm was re-equipping its service with the Boeing 314 clippers. Concurrently the newcomer was having three even more powerful flying boats, Sikorsky V44s, built. After Pearl Harbour, the Sikorskys were immediately taken over by the US Navy, who, like their two rival airlines, were allowed to blatantly ignore Ireland's 'neutrality', carrying warlike stores clearly labelled as such!

Sikorsky Excalibur I over Foynes, in which Captain Charlie Blair made the first Europe to New York non-stop flight in 1942, with just ninety gallons of fuel to spare.

American Export scored a notable first when, on a cool June evening in 1942, the first non-stop passenger flight from Europe to New York took off from Foynes in *Excalibur*. At the controls was an outstanding pilot, Charlie Blair, better known perhaps as the husband of Irish film star Maureen O'Hara. Sixteen VIPs occupied his aircraft's spacious and luxurious cabin, from where they could see several fiercely burning cargo vessels in mid-ocean, victims of the U-boats.

The flight plan called for a stop at Botwood in Newfoundland, about fourteen hours away. However, a signal was received saying that Botwood was 'socked in', with a rough wind blowing drizzle and fog, as were Gander and two further refuelling points. Headwinds had now reduced the Sikorsky's air speed by half, so Blair took it down through the leaden clouds to mast-head level – indeed, the possibility of such obstructions kept the crew on their toes.

The chief flight engineer, an Irishman named Mike Doyle, became quite concerned about their fuel supply. However, by carefully monitoring consumption and with the pilot using 'surface effect' – the cushion of air at low altitude between the wings and the surface of the sea which saves on power output – *Excalibur* made it all the way into New

York's marine terminal at La Guardia Airport. With typical military understatement, one of the VIPs, Admiral Cunningham, crisply commented: 'Remarkable voyage!' Little did he know that only a tiny fraction of fuel remained, after a journey which had started from Shannon twenty-five hours and forty minutes earlier. This *was* a remarkable voyage – the first direct flight between Europe and New York. From an Irish point of view, it had joined 'the auld country' with the city that had taken so many emigrants down the years.

In 1944, Charlie Blair broke the record for fastest Atlantic crossing by flying boat five consecutive times. He continued transatlantic flying in landplanes, having conceived new navigation systems. After the War he set up an inter-island operation, Antilles Air Boats. In one of his flying boats he returned with his wife Maureen to now-abandoned Foynes, thirty-one years after he had made the last official flight from there. Sadly, two years later, he was killed when alighting on the open sea. He certainly deserved the commemorative stamp issued by An Post and the monument to his memory in Foynes Yacht Club.

'Their ocean god was Mananann Mac Lir,' wrote GK Chesterton. This bust in Foynes Yacht Club commemorates the exploits of the great ocean flier, Charlie Blair. It was unveiled by his widow, Maureen.

XIX

MAGNIFICENT MEN AND THEIR FLYING MACHINES

WHEN THE SENIOR LINE SERVICE ENGINEER OF AER LINGUS, who had spent a lifetime in the aviation industry, was clearing his desk on the day of his retirement, and looking forward to cultivating roses at ground level, his phone rang. On the line was a producer from Twentieth Century Fox Films who was planning a Great War aviation epic in Ireland and wanted a chief technical liaison adviser. The producer came out to Dublin Airport to talk to the man in question, Johnny Maher, who gave the proposal an emphatic 'Yes'. As the two journeyed into the city a contract was signed in the taxi. Johnny then joined the airline's General Manager for a farewell lunch over which the host remarked that the veteran technician would surely be sorry to leave the aviation scene. By way of reply the film contract was placed on the table, to the astonishment and delight of the GM. The movie got its title, *The Blue Max*, from the coveted German gallantry award. The lead was played by George Peppard (who actually

did some piloting in the film) opposite Ursula Andress. It marked the beginning of twelve unexpected years in aviation movies for Johnny Maher. But to aviation enthusiasts the real stars were the flying replicas of World War I aircraft, and the real romance was the sight and sound of these early birds reviving the air war over the Western Front.

A line-up of Fokker tri and biplane replicas for The Blue Max.

Ireland was ideal for the project – in County Wicklow there are areas strikingly similar to the landscape of the Somme; its skies are free of heavy air traffic and industrial smog. The movie makers agreed with Shaw that there were no skies like Irish skies. Added to this, there was unlimited co-operation from the Air Corps and Army. A dozen Air Corps pilots were delighted to fly the veteran biplanes in spectacular free-for-all dogfights. The dogfight sequences were painstakingly organised,

choreographed like a ballet. Thus there were no mid-air incidents, even when a dozen aircraft were milling around in a limited block of airspace. (in Hollywood's early epic, *Hell's Angels*, aerial free-for-alls resulted in at least two mid-air collisions.) The realism required

Left: The camera 'copter closes on one of the German replicas.

Off over the Irish countryside goes a gaggle of 'extras', Tiger Moths tricked out in German colours to supplement the exact replicas used in close-up work.

when aircraft were shot down in flames and smoke called for skilful flying by the 'casualties', which came across convincingly on the screen. Unfortunately, the hazards of film flying were demonstrated in a later Irish-based film, with actual casualties being suffered when a warplane and a camera helicopter collided in mid-air.

A Moreine Saulnier parasol monoplane – the latest thing – in which the hero of The Blue Max was to perish – but only in the film!

At Baldonnel, the fleet of twenty-five veteran machines used in *The Blue Max* was housed in an Air Corps hangar. All had been built from scratch, closely following the original structural specifications. Externally, it required a very hard look to discover discrepancies between replicas and originals. This authenticity was confirmed when the aircraft were flown: the handling properties of the clones – good and bad – matched those of their predecessors. The only concessions to modern practice were the provision of radios and parachutes, though, as Johnny Maher remarked: 'A

parachute isn't much good if the plane is operating below 200 feet.' In some ground-strafing sequences, the aircraft were flown below fifteen feet and had to be positioned exactly for the ground cameras – like actors 'picking up their marks' on a set, only more so!

SE 5a replica fighters over the Naas Road.

After battle, weary airmen return in Darlin' Lily.

The German Fokker DVIIs were built in France and the manufacturer's pilot proved their reliability by ferrying each one across from Dinard to Dublin in three hops. They caused many 'double-takes' en route, when the tiny biplanes with their dazzling pattern of brown/black/grey/white lozenge camouflage landed amid towering jet aircraft. Initially the replicas gave serious headaches to the Air Corps mechanics, because they were working on virtual prototypes for which no manuals existed. A fictitious German Staffel (squadron) was based at Weston Aerodrome in County Kildare, which looked very convincing with its existing old farm buildings as squadron headquarters and a row of specially-built canvas hangars. There, the unorthodox Fokker triplanes were proving difficult to land safely because of the extra middle wing, so the experience of an old German pilot who had flown in the great Richtofen's Staffel was called upon. He poured out a long explanation in German, and the film makers were sure that this was a full answer to the problem. At last the interpreter turned to them and summarised the advice: 'He says they used to have a lot of crashes on landing!'

Darlin' Lily: out of control, a wingless plane causes panic on the aerodrome!

The Blue Max production was the beginning of a number of aviation feature films to be made in Ireland – including *Darlin' Lily, Zeppelin, Richtofen and Brown, Aces High* and others. All of these benefited the Irish economy and, of course, gave our Air Corps pilots a chance for some unconventional aerial antics.

XX

CHOPPERS!

IN THE WINTER OF 1954, a fierce hurricane in the Irish Sea caused a large oil tanker to break in two off the Wexford coast. Its crew were marooned on the forward section, which was being driven in towards the Welsh coast. With great skill, the crew of the Rosslare lifeboat succeeded in rescuing all hands, but it was obvious that helicopters should have been on hand to help the brave lifeboat men. The government was approached, but nothing happened. Later, two other sea disasters involved fatalities which would have been greater had it not been for the help of RAF helicopters. Still nothing happened at government level, despite the urging of the media, the fishing industry and the medical profession, all of

All eight Allouettes, including the first three acquired in 1963, still giving excellent service within their limitations.

whom had been agitating for 'choppers' for years. Heavy snowfalls in the winter of 1962 created immense problems in getting severely sick and injured people to hospital. This was the catalyst for government action. At last the Air Corps was instructed to set up a helicopter unit, which was operational, equipped with three French Alouette Mark IIIs, twelve months later.

The Alouette is a highly efficient and reliable helicopter, whose Air Corps numbers eventually rose to eight – the original three are still giving

These conditions are mild compared to some situations with which Air Corps SAR helicopters have to contend.

yeoman service to this very day. Initially, urgent air ambulance missions predominated, giving quick access to accident sites and more tranquil transportation to hospitals. Lives began to be saved. Occasionally these missions of mercy had a funny side – on one occasion, a pregnant woman, already the mother of several children, had to be carried from her island home across very rough seas to a mainland hospital. When the helicopter touched down at the hospital, the grateful passenger pressed a pound note on the pilot who firmly declined the 'fare' and attempted to stuff the note back into the woman's handbag. Unfortunately, just at that moment, a TV cameraman zoomed in and recorded the action, which looked as if the pilot was raiding the handbag! The latter jumped out, got the cameraman by the throat, and warned him of dire consequences should that particular piece of footage ever be screened!

Chopper operations increased – sea-air rescue (SAR) was at last available; Army co-operation and photography were enhanced; and the burgeoning troubles in Northern Ireland caused helicopters to be permanently deployed around the border. The New Year of 1982 was ushered in by continuous snowfalls, far worse than those of twenty years earlier. The chopper squadron was stretched to its limit. Requests for help came in from every quarter, for by now 'Air Corps Helicopters' was a household phrase – though only a daylight service was on offer. The Alouettes are not equipped for night flying, though two stranded climbers were rescued one night from a treacherous mountain face by one chopper, using the landing lights of another to illuminate the scene. The pilot of the former was deservedly awarded the first of many Distinguished Service Medals to be won by men of his unit.

Disaster at Whiddy oil refinery when a tanker blew up! Air Corps Alouettes were in the forefront of the rescue endeavours.

By now it was becoming clear that training fixed-wing air crew on the Alouettes was eating up valuable operational time, so a couple of nimble single-turbine machines were acquired from the same manufacturers. Still, more operational choppers were needed. Again, from the same stable, five Dauphins were purchased, two of them specially equipped to work with the Naval Service on SAR and marine protection duties. But even today the Air Corps lacks the medium-range, medium-lift helicopters which could cover the country's territorial responsibilities properly. Conditions around our shores vary immensely –

from the east coast, where waves rarely top three metres, to the mountainous Atlantic seas, which can reach twelve metres. To cover the Atlantic properly a civilian contract has had to be arranged, hopefully as an interim measure, whereby large Sikorsky helicopters are based at Shannon. The Air Corps covers the southeast with Dauphins deployed at Waterford Airport. The dangerous nature of the Squadron's duties was tragically highlighted in 1999 when, out searching for a missing boat in dense fog, a helicopter collided with high ground, killing its crew of four.

Irish Helicopters was formed in 1968. It got its first break when Irish Lights contracted the new company for year-round relief and resupply of its manned lighthouses, until the last one became automated. Six years later Aer Lingus bought a majority interest and soon became the sole owner. The choppers were, and still are, also involved in support work for offshore oil drilling and development. As this exploration expands helicopters support the rigs by transporting men and equipment to and from the mainland. Irish Helicopters's most notable VIP passenger was Pope John Paul II during his 1979 visit, though others have included royalty, presidents and prime ministers. In 1985 Irish Helicopters began to operate a fortnightly service to Tory Island, which is frequently cut off from the Donegal mainland by the weather.

'You're not going home dressed like that?'
Two helicopter crewmen at the end of an operation.

As the Air Corps's usage of helicopters has expanded, other organisations have been drawn to the non-military capabilities of the chopper. The geographical make-up of this country makes their

A medium-lift Puma helicopter, which the Air Corps tried out on a year's lease.
This is the type that is badly needed now if the Corps is to fulfil its SAR commitments.

employment particularly important: choppers are vital when access is needed to remote locations. The film industry was quick to use such a versatile camera platform, and early on the Air Corps earned useful revenue providing this service. In many movies, a hovering chopper permits dramatic and otherwise unobtainable camera angles. Helicopters are increasingly used in such diverse occupations as the tourist trade and the building industry – in fact, in any situation requiring access to a difficult location the rotary wing can easily outfly the fixed wing. And to think that Leonardo de Vinci, artist and engineer, had it all figured out in 1483!

XXI

FROM GREEN TO
TWO SHADES OF BLUE

AT THE END OF WORLD WAR II, or the 'Emergency' as this period was known in Ireland, the Air Corps was equipped with somewhat war-weary Hawker Hurricanes. These were replaced by a full squadron of Supermarine Seafires, the naval version of the famous Spitfire. As the Cold War 'hotted up', the squadron undertook an important assignment, known as Operation Sandstone. Ireland was, in effect, the seaward flank of the free world, vulnerable to sea-borne landings. Though the country

Seafires, which, operating on single sorties, mapped out Ireland's coast in Operation Sandstone during the 1950s.

had been invited to join the newly-formed North Atlantic Treaty Organisation (NATO), it declined, partly because this would involve agreeing to all existing frontiers, including the northern Border. Nevertheless, the US and UK authorities were keen to have a thorough photographic survey of the coast and its hinterland. Concurrent with this, an assessment of possible radar sites and the effects of coastal erosion were domestic requirements.

The survey begun in 1948, the year NATO was set up, and took about five years, this timespan being dictated by weather and aircraft serviceability. It called for highly accurate flying in terms of altitude and speed; the Seafires' cameras had to be carefully sighted via calibrations on cockpit canopies and wings. The heavily indented coastline meant intensive sorties of up to ninety minutes' duration to take advantage of clear spells. The British contribution was seen in the provision of a helicopter, a type of aircraft which was then relatively new. When it arrived at Finner Camp, the local Donegal people did not welcome its presence. Tensions were exacerbated when the corrugated buildings of the base, which had been green for many years, received a badly needed coat of *red* paint which the *Donegal Democrat* interpreted as further British

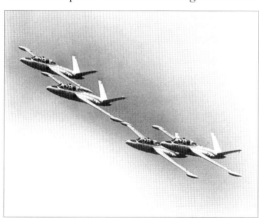

The Air Corps's 'Silver Swallows' acrobatic team, now, alas, disbanded as these Fouga Magisters are retired.

influence at the base! The helicopter, incidentally, because of its short range, had to be transported around the country on a low-loader and eventually met its end by crashing into the graveyard at Youghal. Luckily the crew did not join the residents there!

The Air Corps joined the jet set in 1956 when DH Vampires were acquired. These were subsequently replaced by Fouga CM170 Magisters in 1975. The latter formed an aerobatic team called the Silver Swallows, which was to show the flag dramatically and win awards at air shows at home and abroad.

Instructor and pupil in a SIAI-Marchetti SF 260 WE Warrior, a type which has proved to be a versatile light strike, gunnery training and basic flying training aircraft.

In addition to operating the high-profile helicopters and the government jet, which occasionally becomes a political football in terms of operating costs and utilisation, the Air Corps undertakes intensive maritime surveillance. This covers not only territorial waters, but also, since the country joined the EU, the European Economic Zone (EEZ), which stretches 200 miles out into the North Atlantic. A pair of CASA aircraft targets illicit fishing boats, drug and arms smugglers and oil slicks, which are duly reported to the ships of the Naval Service. The airmen's long patrols are occasionally enlivened by sightings of whales. These massive creatures can be very plentiful when those from southern waters join with their northern cousins. But there is a need to cover the holes in the surveillance net, which requires additional aircraft.

The new millennium poses new problems for the Air Corps, as well as for the rest of the Defence Forces. Many factors have to be addressed, including the drain of pilots and technicians transferring to more lucrative jobs in civil aviation. The Corps's personnel have got the skills, dedication and professionalism to carry out

The government jet, a Gulfstream IV, with the Presidential crest on its tail fin.

the tasks allocated to them. Their motto, *Forfaire agus Tairseact* ('Vigilant and Loyal'), is no empty slogan. An independent report has recommended that four medium-range, medium-lift helicopters are urgently needed but, in fact, for comprehensive SAR (sea-air rescue) cover, a total of six, complete with state-of-the-art navigation systems, are necessary. Their acquisition seems to be a leisurely process. Air cover over the EEZ calls for at least four more CASA aircraft to provide a twenty-four-hour, all-weather service and it makes sense to base them, and some SAR helicopters, at Shannon, close to their area of operations, instead of at Baldonnel, 200 miles away to the northeast.

The CASA is king! Having completed 10,000 hours of trouble-free operations with this aircraft, the Air Corps invited other users to take part in a competition which involved filing a flight-plan including an ETA (estimated time of arrival) at Baldonnel. The prize: the winning captain's own weight in Jameson's golden product! The winner, a captain from the Spanish

The CASA has a six-man crew – two pilots, two radar and sensor operators, a photographer and launch operator – whose spacious surroundings give them the requisite comfort on patrols of up to ten hours' duration. When on SAR missions, the crew can drop life rafts and emergence rations through the rear ramp, as well as various types of markers. The Maritime Squadron not only provides fishery protection patrols, but also keeps an eye on the Economic Zone, an area of 132,000 square miles.

Air Force, weighed in at six cases, one bottle and a *taosgán!*

The downsizing of Ireland's UN peacekeeping military contribution is balanced by its commitment to a European Rapid Response Force. A mechanised infantry battalion of up to 1,000 men has been promised for this force's 60,000 strength. Undoubtedly, in time, a credible airforce will be required for our EU Atlantic border, and Ireland will no longer be able to shift this responsibility onto its partners, nor hide its head in the sand regarding domestic sovereignty and security. Currently, as the ancient Fouga jets retire, there is a requirement for at least eight single-seat strike aircraft with suitable trainers. The 'Celtic Tiger' is reckoned to be cash-rich, and surely some of this prosperity should allow the claws of the Black Panther crest of the old First Fighter

The Eurocopter 'Squirrel' Garda helicopter over Dublin.

Squadron – no longer on the inventory of the Corps – to be sharpened?

The Air Corps changed its traditional green uniform to one of blue in 1944. Soon after, a lighter shade of blue was seen around Baldonnel with the inauguration of the Garda Air Support Unit (GASU). Like the Air Corps helicopters, it was a long time coming.

On an April day in 1934 a thief raided a shop in the village of Clondalkin and beat up its owner. When the alarm was raised the local Garda Sergeant sought the help of the Irish Aero Club's Chief Flying Instructor at nearby Baldonnel, and both took to the air in a search for the miscreant. Flying low over fields and lanes they spotted their quarry and the CFI landed close by in a field. The sergeant gave chase, but didn't catch the fugitive. However, as the newspapers reported: 'This demonstration of the utility of aircraft should not go unheeded by the government and, though the pursuit was unsuccessful in this instance, it helped to show how aeroplanes can assist the police in their work.'

Garda Defender in flight.

A few years earlier, police forces in many countries were experimenting with fixed-wing aircraft, auto gyros (the forerunner of the helicopter) and even airships. But it is only during the last decade or so that the concept of airborne law enforcement has been fully accepted. The various police forces in the UK tend to use twin-turbine helicopters, mounting forward-looking infra-red (FLIR) and TV cameras, microwave downlinks and other aids, but experience has shown that fixed-wing aircraft also have their place. It is not a matter of 'either/or'; rather it is the realisation that both types of aerial platform have specific advantages and each must be considered as complementary to the other. This is the combination the Garda opted for. Earlier, the spillover of the 'Troubles' into the Republic had caused the Gardaí to rely on the Air Corps. The crews would usually consist of a pilot and a Garda without any air observer training. Communication between the policeman and the ground was virtually non-existent, principally because he would not possess the necessary map-reading skills.

In addition to the Border patrols starting in 1972, putting a 'copper in a chopper' included monitoring cash-carrying vehicles in case of ambush. From 1990, police surveillance from fixed-wing aircraft began, but there were problems. The military aircraft were not adapted for the Garda operational role and there was none of the special equipment necessary for police operations. In addition, the work was dependent on daylight and good weather. Still, these experiences did point the way for future improvements, and the Gardaí became more air-minded. The major catalyst for change occurred in 1980, when two Gardaí were killed during a bank robbery. A fact-finding mission throughout Europe was undertaken and the commissioner of the force strongly urged that his men needed dedicated air support. At last, in 1994, an air support service was established, not only because of the political situation in Northern Ireland but because drug-related crime and gangland killings were on the increase in Ireland as elsewhere. The plan that emerged was that the Air Corps would be responsible for piloting and maintaining police aircraft, but suitably skilled personnel of the Garda Air Support Unit (GASU) would form the crew.

The first two aircraft were built. They were a twin-engine PBN Defender 4000 aeroplane and a Eurocopter Ecureuil (Squirrel) AS355N helicopter. All police units throughout the country were alerted to the presence of their new airborne colleagues, whose mission was clearly stated as: 'To provide a patrol/response/surveillance capability, both overt and covert, in the assistance of operational ground units of An Garda Síochána'. The new aircraft were equipped with basic equipment in order of priority, and 'like to have' options – such as moving map display, simulcast and spotlighting – were considered. The aircraft were given Air Corps serials – the Defender bore the number 254 and the helicopter 255. Eventually, on 1 September 1997, the unit was declared operational, and on the seventy-fifth anniversary of the founding of An Garda Síochána the Commissioner was helicoptered out from his Dublin office to Baldonnel, where the justice minister declared that the Garda Air Support Unit was 'open for business'.

As the criminal underworld tends to operate in darkness, the Garda operate FLIR cameras, which can turn night into day with a clear picture. During daylight hours they can locate fugitives by sensing their body heat, even when they are concealed. The Defender is much quieter than the helicopter and its ability to fly and manoeuvre at very low speeds gives it a stealth capability.

Of course the predominance of the helicopter in police fleets will never be challenged by the fixed-wing aircraft; however, the latter has many advantages, not least of which is its cost-effectiveness compared with the helicopter.

The results of this major development from the 'copper on the beat – or in the car' was simply phenomenal and now another helicopter, a Eurocopter PC 135, capable of taking an eight-man crew, has been added to the fleet. No doubt this new addition will increase the GASU personnel from sixteen, plus a pool of ten pilots. By 1999, over 1,600 hours had been flown, resulting in 520 arrests, fourteen missing persons located and 208 stolen cars recovered. One technique that has been found to be very effective is a 'downlink', whereby live images can be transmitted back to Gardaí on the ground. The 'air bears', as American truck drivers say, have also made a great difference in keeping traffic moving, as local authorities now have a first-class information service for road layout and traffic sequencing.

It hasn't all been smooth sailing. There was bound to be a 'culture clash' between the ways of the policeman and those of the military. This was only natural. For example, the seating configuration of the Squirrel helicopter meant that the views, literally, of the pilot and the Garda observers conflicted. The pilot's view to the left was blocked by burly policemen! A compromise was reached when the sliding door with its 'picture window' was changed from port to starboard, leaving everybody happy.

The Garda Squirrel is one of an ubiquitous species! There are more than 150 being used by police forces in over twenty countries worldwide. It has excellent cockpit visibility, can turn 'on a sixpence', and can get to the scene of a crime at high speed. If a police unit is needed on the ground it can deposit four officers on the spot. In gentler mode, the Squirrel is used for monitoring road traffic, including transmitting images back to a local control centre. It can be used to protect the environment by detecting and preventing forest fires and, if necessary, can take an intervention team of firemen to the scene of a fire.

XXII

THE CIVIL SCENE

AER LINGUS'S HECTIC POST-WAR ACTIVITY proved to be 'too much, too fast', and the airline was losing money, to the tune of £650,000 in 1947/48 – equivalent to £15 million in today's money! Nevertheless it was looking to the transatlantic route as a prospective money-spinner. To tap into this market a subsidiary company, Aerlínte Éireann, was formed, but an incoming coalition government maintained that the country simply could not afford the investment needed. Commitments already made had to be scrapped – five Lockheed Constellations had been delivered, and Aer Lingus crews, who had transferred to the new subsidiary and had already been trained on them, were made redundant. Luckily many secured positions with BOAC, which was also very happy to buy the Constellations. Not so happy was half of a class of twenty Air Corps trainee pilots who, after military service, would have been able to transfer to transatlantic flying – ten cadets had to go and find their feet in other careers.

The blow fell when Civil Service department heads met the new Cabinet. One of them, who was also Chairman of Aer Lingus, courteously invited the Minister for Finance to participate in the first transatlantic flight, scheduled for St Patrick's Day. The Minister replied firmly: 'There will *not* be a first flight.' Before Aer Lingus disposed of the Constellations, it operated them for a short period as a 'Silver Shamrock Service' to London and Rome, routes for which these sleek, shark-like aircraft were uneconomical.

A decade later, Aerlínte finally got off the ground – but not in its own right. The company leased Super Constellations from Seaboard and Western Airlines, a charter cargo airline already using Shannon. The aircrews were American, but the 'Connies' were in Aerlínte livery and the cabins staffed by one of the company's greatest assets – the friendly Irish girls! Better late than never? Some believed that Aerlínte Éireann had lost a golden opportunity by not accepting the idea when it was first proposed. The Irish manager of TWA commented on the ten-year delay:

After a ten-year delay, Aerlínte Eireann finally operated an Atlantic service using the superb Lockheed Constellations.

They were the great years. Aerlínte would have been one of the first transatlantic carriers – it would have been in before many European national carriers, and they could have built up valuable experience over that ten-year period.

But Garrett Fitzgerald, who had worked in Aer Lingus, disagreed:

Closing was absolutely the right decision at the time. It was opened at the right time ten years later, when the traffic was building up to the point where it was possible to envisage an economic operation.

Whoever was right, by 1960 Aerlínte was operating its own transatlantic aircraft. Both European and Atlantic fleets now sported an all-embracing

1947: An Aer Lingus air hostess, an important member of what was then known as 'The Friendly Airline'.

green shamrock and the legend 'Aer Lingus – Irish International'. Later, this was shortened to 'Aer Lingus'. A livery change featured a *white* shamrock, but this hybrid did not survive for very long.

For European routes, a fleet of Vickers Vikings, a civil type based on the RAF's Wellington bomber, was introduced, but these were heavy on fuel and generally unsuitable, so they were sold on. In 1954 a major, courageous, decision was made – the airline converted from piston-engined aircraft to an all turbo-prop fleet. The four-engined Vickers Viscount was introduced but initially proved troublesome, particularly when, after four years in service, the manufacturer advised that the wings be modified! Later, a newer model proved far superior and extremely profitable.

In his autobiography, *All in a Life*, Garrett Fitzgerald recalls one marketing stroke brought about by Marian devotion in Ireland:

> The Aer Lingus predominance at Lourdes reflected, I had reason to believe, an irrational prejudice among some other airline managements against what they seemed to feel to be superstition: a feeling that made several airlines act perversely and non-commercially so far as Lourdes was concerned. Even our British shareholders up to 1957, BEA, showed hostility to our Lourdes operations, apparently suspecting that we were running these services for religious rather than commercial reasons, a belief that I found quite entertaining.

> In any event, with the centenary year coming up, I asked Stuart-Shaw [the General Manager] to defer until autumn 1958 the sale of two of our Viscount 700 aircraft, which were being replaced with the larger and more economic Viscount 800 series. I told him I could use them profitably for Lourdes charters during the five months from May to October. He was sceptical: even *he* seemed to suspect me of a non-commercial, religious motivation! I persuaded him to agree,

however, and as a result we achieved what I still believe to have been the most perfect supply-demand equation ever recorded in air transport. The two aircraft were used on that route twenty hours a day on *every* day of that five-month period, and we secured eighty-one of the eighty-two pilgrimages from Ireland and Scotland to Lourdes that summer.

Complementing the Viscounts, Fokker Friendships later arrived from Holland, Aer Lingus being the second customer, after KLM, the Dutch national airline. These aircraft were also profitable on short-haul routes. Their twin engines, Rolls-Royce Darts, matched those of the Viscounts, thus standardising maintenance. The order for the Friendships came as 'the friendly airline' celebrated its twenty-first birthday – with a flypast around Dublin led by a demonstration Friendship, flanked by two Viscounts and followed by a flight of six good

The DC3 fleet builds up.

old reliable DC3s. As the latter returned to the their base, they broke formation in a spectacular bomb-burst over the airport. Apart from some initial problems which the makers soon put right, the forty-seater, high-wing, turbo-prop Friendship entered service with unusually few teething problems, and the type was a favourite with pilots and customers alike.

A formation of six Aer Lingus Douglas DC3s, led by Captain Ivan Hammond, breaking up in a classic 'bomb-burst' over Dublin Airport after the twenty-first birthday fly-past over Dublin.

BAC One-elevens ushered the airline into the short-haul jet age in 1965.

Aer Lingus entered the jet age in 1964 with Boeing 707 320Cs, and the next year acquired four new BAC 1-11s. Although Friendships were still in the air, there was nothing like friendship on the ground. The company's staff were looking for more money and better conditions. There were actual and threatened strikes, but out of all this unfriendliness emerged a funded pension scheme for the company. It became the first state enterprise to have such an arrangement, which the staff felt offered greater security of employment.

At the beginning of the 1980s Aer Lingus took a hard look at its 'thinner' routes, those between Dublin and provincial British cities. The airline had the choice of pulling out of these uneconomic routes or approaching them from a totally new angle. Management and pilots calculated that, with much smaller aircraft and new changes in working practice, the routes could be made profitable. Thus it was that thirty-seater Short's 330s, built in Belfast, were introduced in 1983 in a new Commuter Service. The Chief Executive expressed his feelings at its inauguration: 'One of the things which has given us a lot of pleasure is that for the first time we are flying aircraft which are actually manufactured in Ireland.' The first of the type was named *St Ultan* and registered EI-BEG. However, lest it appear that Aer Lingus was 'begging' for passengers, somebody decided to advance the final letter of the registration to 'H'. Many more Belfast-built aircraft were to come on-stream as the basic design was developed.

But there were now other Irish carriers. An 'upstart' was taking a slice from the national carrier's twenty-first birthday cake! Ryanair, which describes itself as a 'no-frills' airline, took off in June 1985 from the newly-opened Waterford Airport for Gatwick with a lone Brazilian fifteen-seat Embraer Banderiante. Soon, 'the other Irish airline' was very busy on British routes, but this did not stop the newcomer from simultaneously introducing a charter programme, flying holidaymakers to sixty-five different European destinations.

With the shamrock well-established as the Aer Lingus logo, Ryanair opted for the other national symbol.

By the end of the summer of 1987, Ryanair had carried a quarter of a million passengers. The European Community declared that government-protected national airlines must now stand on their own feet, and Ryanair broke into the tightly-guarded and lucrative Dublin/London route. The impact was dramatic – those who said that the route was already saturated, and that the result would be aircraft flying half-empty, were confounded. In fact, traffic increased and all carriers on the route benefited; the Irish at home and in Britain were quick to appreciate that competition is truly the life of business. Since it first took off more than fifteen years ago, Ryanair has spread its wings far and wide, both domestically and internationally.

Is there a downside to the Ryanair success story? Well, other low-cost airlines have come and gone. Ryanair says of some of its competitors that they are flying to too many and too expensive airports. Of British Airways subsidiary 'Go', Ryanair maintains that BA is trying to change overnight its long-standing philosophy of looking for profits while abhorring 'cross-functional' staff. 'Anyone can do anything' is the philosophy that seems to make a 'no-frills' approach work. But now the 'Go' subsidiary has gone out on its own, free from BA's apron strings. Does the low-cost philosophy work? One can only wait and see; certainly when companies have tried to compete against entrenched flag carriers by offering lower fares, many have come unstuck. Whether travellers will always be happy to use far-out-of-town airports, to fumble for money for coffee, to forego in-flight meals and, in the case of Ryanair, to pay for the privilege of using credit cards when booking, remains to be seen. In any event, David has taken on Goliath – and the travelling public and shareholders will watch the contest with interest.

Worries, both worldwide and domestic, as well as wars and recessions, brought Aer Lingus to the brink of bankruptcy in the early 1990s, but the company survived. At the beginning of 1996 it embarked on its 'Strategy for the Future' when it stated that it was investing for the new millennium. Part of this strategy was a new 'corporate image', to underline that the airline had soldiered on in dark days and was now initiating a radical restructuring programme. To the ordinary punter the immediate sign of change is the new appearance of its aircraft, both inside and out. The chameleon may blend into its background, but a dynamic airline must stand out from its rivals. It must display its wares in all possible ways – in its advertising, in its ticketing, timetables and brochures, and of course it must make its aircraft stand out when parked on airport aprons.

Thus, Aer Lingus initiated an £8 million visual scheme which it had been mulling over for years. The 'hard years' had proved that history and heritage alone matter little unless the goods are clearly perceived in the marketplace. The Aer Lingus rethink took on board the importance of colour and design to uplift the image and heighten the perception of Ireland and the Irish people internationally. The central concept was 'a celebration of the vibrancy of our literary heritage'. Our long list of novelists, poets, and musicians is widely recognised – now, travelling on an Aer Lingus plane, one can read their work woven into the upholstery of the seat in front!

Though no amount of gloss on a plane will hide low standards, when they are high they must be flaunted, the flag must be waved! The shamrock is retained because Aer Lingus believes that this symbol has long signalled the company's trustworthiness, competence and reliability. Everything is now presented in shades of green, blue and grey, 'to evoke the lush and verdant landscapes interspersed with clean lakes and rivers and overcast with mist-laden skies'. Sheer poetry – but the overall marketing strategy seems to be working in the hard-nosed business of transporting passengers and cargo for a realistic profit. Recently the company became the ninth member of the 'Oneworld Alliance', meaning that its customers can transfer 'seamlessly' to its partners' routes – though already the Irish line's own network covers routes from Ireland to eight destinations in Europe and to Boston, Chicago, Newark, and Los Angeles.

We have moved into a new millennium, with the promise of more technical wonders for airlines and air forces all over the world. Travel by air will increase steadily, and the military scene will change too – designers are already working on the first unmanned fighter plane. Shuttle planes are already being boosted into the stratosphere to look at other planets and then return to Earth, landing on a conventional runway like a conventional aircraft.

What are the implications for the future? Will there be bigger fixed-wing planes and larger helicopters, or a combination of both? Aircraft capable of taking 1,000 or more passengers, the population of a village, are already being designed. Will this decrease the ever-growing frequency of flights? Will the development of fan-jet engines mean quieter skies? Will the Irish sky become clearer, and less polluted by noise and fumes? Will more green land be given over to more concrete runways and terminals? Time alone will tell. No doubt the romance of flight will always entice human beings – as it did long ago when a poet penned his hopeful view at take-off:

> Forth to the clear of the runway gleaming
> And gashed with gold where the grey lies wet,
> And the line-stones white on the turf go streaming,
> Fast, and faster, and faster yet.
> 'Ere the plane leaps up like a barb of fire,
> To its utter achievement of winged desire –
> And croft, and cottage, and hall and spire
> Grow dim as a dream of deep regret.

In about four years, the Airbus Industrie's A3XX–100 will take to the skies. There is nothing nothing in this photograph with which to compare this giant, but with a capacity of 555 passengers in a typical three-class configuration, it is 260 feet in length against the Boeing 747's 232 feet. The range of the Airbus A3XX with a full load is 7,165 miles. The 100 Series will utilise the latest state-of-the-art technical developments to provide lower seat–mile costs and satisfy environmental considerations. It will become 'the baseline for the future A3XX family'.